FIREPOWER
MECHANISED WARFARE

FIREPOWER

MECHANISED WARFARE

TACTICAL ILLUSTRATIONS • PERFORMANCE SPECIFICATIONS
FIRST-HAND MISSION REPORTS

EDITOR: CHRIS BISHOP

Grange BOOKS

Contents

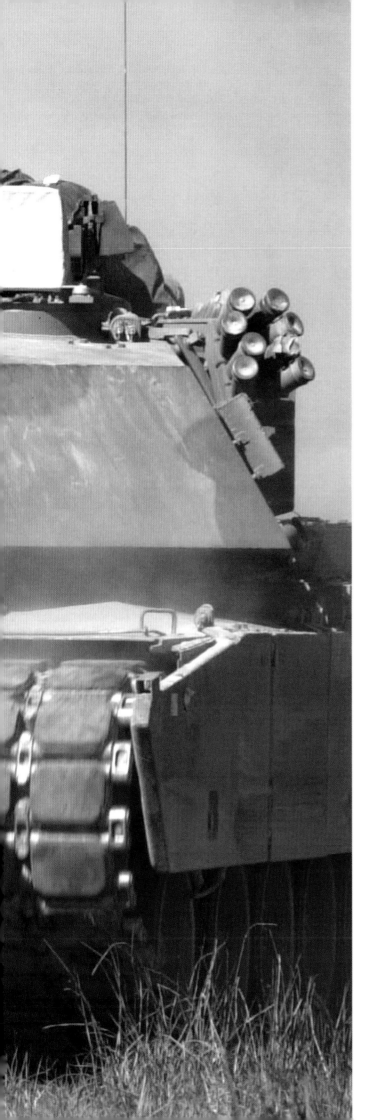

Introduction

The beginnings of mechanized warfare were quite inauspicious. In September 1916, some 50 British Mk I tanks rolled slowly into action during the Battle of Flers-Courcelette. Mechanical reliability was so poor that 18 had broken down even before the start orders were given, and the rest were almost impossible to manoeuvre, let alone fight in.

Yet despite the undoubted lack of decision they imposed upon the WWI battlefield, the Mk I tanks set in motion a trend towards mechanized warfare that is still being advanced today: armoured vehicles offer a potent trinity of protection, firepower and mobility in one. It took another 20 years before technology and tactics were able to give full expression to this vision. In 1939 and 1940 Germany launched its *blitzkreig* into the heart of Europe, demonstrating the awesome power of combined-arms operations. The vehicles which emerged from WWII actually laid the foundations for the modern Main Battle Tanks we see today. Furthermore, armoured vehicles shifted into a new spectrum of roles, from reconnaissance and mine-clearing to bridge-laying and communications.

The lesson of WWII was clear – full mechanization of armies was essential to achieve dominance over the enemy. The Cold War gave further weight to these lessons, as the superpowers prepared for possible massed motorized warfare on the plains of Central Europe. Yet while the actual roles of military vehicles changed little in the post-war period, what has changed is the awesome progression of technology. From the Mk I's ineffectual six-pounder guns we now have systems like the Multiple Launch Rocket System which can devastate hundreds of metres of ground in a single strike.

The Gulf War in 1991 proved that mechanized warfare still has a place in the post Cold War world. Yet on the other side, the technologies to destroy armoured vehicles have also kept pace. A pair of today's infantrymen can deploy the weapons to take on and destroy even the largest tank. The challenge for today's mechanized armies is as much about survival as about dominating the battlefield.

Firepower: Mechanized Warfare provides a detailed historical and tactical insight into the practice of mechanized warfare. Not only does it explore the machines and strategies which have defined motorized combat, but it also gives first-hand accounts of what it's like to fight in armoured conflict, past and present.

The M1A1 Abrams MBT (Main Battle Tank) is the world's most advanced armoured fighting vehicle. Here we look straight into the 120-mm Rheinmetall smooth-bore gun, which can fire Armour-Piercing Fin-Stabilized Discarding-Sabot (APFSDS) rounds.

US M1 tanks in the Gulf War utterly decimated the Soviet-built Iraqi T-64 and T-72 machines, which were totally outclassed on target designation and range. In light of the conflict, Russia is completely overhauling its Main Battle Tank design.

BATTLE GROUP

British tank and infantry regiments fight together in combined units called battle groups – the fastest-moving and hardest-hitting type of unit in land warfare.

The forestry block lay in darkness, tall German pines blocking the first rays of dawn sunshine. Nothing stirred. Suddenly, the silence was shattered by the deep bellow of a powerful diesel engine. Then another engine roared nearby, and another as the sleeping giants stirred.

Now the whole wood was alive, the ground vibrating; figures scuttled about in the murk, checking over the Challenger tanks and clustering around the back doors of the personnel carriers. The smell of hot tea mingled with exhaust fumes and the powerful odour of fuel oil. In minutes the first tank rumbled forward, its 65-ton bulk squashing a line of young saplings. The battle group was on the move again.

Land combat is still dominated by armoured units. Only they have the fire and mobility to win a decision on the modern battlefield. Since neither tanks nor infantry can be successful on their own, all major armies mix their infantry and armour into combined arms formations, adding artillery and engineer elements to produce a powerful team.

Self-propelled artillery and rocket launchers manoeuvre 10-15 km behind, ready to bombard enemy positions before the battle group attacks. Armoured engineer vehicles travel with the battle group itself, bridging obstacles like anti-tank ditches or minor rivers. If the battle group is on the defensive, they can plant mines and create obstacles for both men and vehicles.

Wide deployment

To the infantryman used to fighting on his feet, the scale of the battle group's operation can be quite a shock. The battle group may move 20-30 km during a single day, operating on a frontage of 5 km or more. Although it includes over 250 vehicles at full strength, they are scattered over a very wide area.

Scouting far ahead of the main body, CVR(T) Scorpions and Scimitars probe the enemy reconnaissance screen to provide the

The British Army's Challenger Main Battle Tank has an imposing 62-tonne bulk. But even this monster cannot fight alone. In a battle group, it works closely with infantry travelling in armoured personnel carriers.

9

The 28 tanks of the battle group can consume enough fuel in a day to drive a family car eight times around the world

battle group commander with an up-to-date picture of enemy strengths and actions. They work closely with Gazelle helicopters, which can zoom across the battlefield to investigate contacts and literally allow the commander to see 'the other side of the hill'. If the Gazelles discover a concentrated mass of enemy tanks, they can vector in Lynx helicopters armed with TOW anti-tank missiles. But, unless the enemy are in overwhelming strength, their armour will be dealt with by the Challenger tanks of the battle group. A modernised version of the long-serving Chieftain, the Challenger has a good cross country performance and its 120-mm gun can knock out any tank currently in service.

The tanks and infantry advance using the same principle as a rifle section on foot: fire and movement. One element advances, covered by the other with 'one foot on the ground' at all times. Infantrymen ride in armoured personnel carriers which alternate between waiting out of sight, tucked behind buildings or trees, and breakneck dashes to the next piece of cover. The 15-ton FV432 lurches and bucks wildly as it crashes over the country at top speed.

Introduced nearly 40 years ago, the FV432 has surrendered most roles to the Warrior. This is designed for much faster cross-country speed – it can happily break the speed limit on a motorway – and has a turret mounting a 30-mm cannon and 7.62-mm Chain Gun. When the infantry dismount to assault the enemy, the objective will be lashed with cannon and machine-gun fire to suppress hostile fire.

There is little room in the back of either type of APC once an eight-man section has piled in with all its kit. After a few moves, the

Battle groups include a handful of MCT (MILAN Compact Turret) vehicles, which fire anti-tank missiles capable of destroying most Soviet-manufactured armoured vehicles. Special sights allow the gunners to engage the enemy through smoke or at night.

troop compartment is thick with dust and oil. The driver takes his orders from the vehicle commander behind him: controlling the vehicle at speed over rough ground is no easy business, especially when operating 'closed down' with the hatches shut. But at least the commander and driver know where they are and can see something of the outside world. For the men crammed into the troop compartment, it can be very disorientating.

As the APCs speed from cover to cover, getting ever closer to their target, the tanks are moving into position ahead. The battle group's command vehicles are tucked away in an area of woodland, monitoring the progress of each unit on the radio. One of these converted FV432s houses the commander of the infantry companies; in another, the artillery commander is in comms with the distant batteries of self-propelled guns to the rear.

Right: A retired American M47 tank is destroyed by a direct hit. The large warheads of anti-tank missiles like MILAN can penetrate up to a metre of conventional armour plate. The only defence is the advanced armour fitted to the British Challenger or American M1.

FLASHBACK

A soldier on foot could outpace a British tank of 1918, but the idea of all-armoured formations had been born.

Forgotten lessons

Tanks were developed during World War I to break the deadlock on the Western Front, where Britain's artillery — the heaviest concentrations in history — had failed to blast a path.

By 1918 the British Army led the world in its use of armour. Lumbering giants like the Mk IV and B tanks crushed their way through defensive positions. Then Whippet light tanks, travelling at a modest 8-10 mph, could pursue, engaging the enemy with their machine-guns. When the war ended, plans were being made to provide similar vehicles to carry infantry: the battle group had arrived.

Unfortunately, although the theory continued to develop after 1918, the British Army displayed less and less interest. Although it was a fully mechanised force by 1939, its tactics remained firmly rooted in the trenches of World War I: the idea of combined forces of tanks and armoured personnel carriers had been forgotten. But in May 1940 the Germans put the theory into practice, and the following year they developed it further: massive combined arms forces such as *Panzergruppe Guderian* plunged deep into the Soviet Union.

A Comet tank in Normandy: not until 1944 did the British Army receive tanks able to match German armour.

BRITISH ARMY BATTLE GROUP Reference File

49

UNITED KINGDOM

Challenger Main Battle Tank

Challenger tanks are the most powerful armoured vehicles in a British battle group. Their 120-mm guns will destroy any tank currently in service while they enjoy the protection of the resilient Chobham armour. This composite material offers maximum protection from infantry anti-tank missiles. Israeli combat experience has confirmed that trained tank crew are harder to replace than the vehicles themselves; hence the concentration on thick armour at the expense of speed and weight. At 62 tonnes, a Challenger is 20 tonnes heavier than typical Soviet-era tanks. It cannot quite match the cross-country performance of a T-64 or T-80. But its 1200-hp diesel makes it remarkably

The Challenger is one of the most powerful tanks in existence but even this 62-tonne monster could be destroyed by a Russian 125-mm tank gun.

nimble for such a heavy vehicle.

Challenger is a development of the Chieftain tanks that had been in service with the British Army since the 1960s. They were designed for the Shah of Iran, but were supplied to the British instead after the 1979 revolution. In the 1990s Challenger received a substantial upgrade, which resulted in the Challenger II with improved armour and a new turret.

Specification
Crew: 4
Combat weight: 62 tonnes
Engine: 12-cylinder diesel developing 1,200 bhp at 2,300 rpm

Top road speed: 60 km/h
Armament: 1×120-mm gun with up to 64 rounds; 2×7.62 mm machine guns with 4,000 rounds; 2×smoke rocket dischargers

50

FV432 Armoured Personnel Carrier

Most British infantry rode into battle on an FV432 before the Warrior was introduced in 1984. Yet some 1600 are still in service with the British Army, though these will go soon when the new MRAV enters service. The FV432 is 40 years old and it shows its age: it is slower than modern American or Russian APCs, and mechanical reliability is becoming a serious headache. But the FV432 has evolved new roles that keep it in use. The command vehicles, self-propelled mortars and other specialist versions will soldier on, sustained by spare parts released from the redundant APCs.

The vehicle's armour is designed to keep out rifle and machine-gun fire and

The FV32 has mostly been replaced as an APC by the Warrior, but its many specialist versions will continue in service. Intended as an armoured transport, it is not designed to engage enemy vehicles.

shell fragments. Many FV432s have been fitted with a turret mounting a single 7.62-mm machine-gun; others have their gun on a pintle mount in front of the commander's hatch.

Specification
Crew: driver, commander and 10 infantry
Combat weight: 15 tonnes
Engine: Rolls-Royce multi-fuel two-stroke six-cylinder developing 240 bhp at 3,750 rpm
Top road speed: 52 km/h
Armament: 1×7.62-mm machine-gun

Components of a Battle Group

Armoured personnel carrier	Carries infantry onto enemy position	
Main battle tank	Destroys the enemy armour by direct fire	
Armoured recce vehicles	The eyes and ears of the battle group	
Self propelled artillery	Indirect fire support to battle group	
Mobile anti-tank weapon	Commander's reserve tank busting weapon	
Recce helicopter	Airborne observation and recce teams	
Infantry anti-tank weapon	Main anti-tank system; static and plentiful	

He calls for a fire mission and separate groups of guns, scattered in concealed positions 10-15 km away, raise their barrels to the sky. Their 155-mm shells contain over 42 kg of high explosive, which will pulverise the enemy position just before the tanks begin their attack.

The APCs come to a jarring halt near the objective. The door bangs open and the infantry hurl themselves out. Surging from the vehicle, they are now at their most vulnerable, but they rapidly deploy into a line, six to eight metres apart.

Divided into two four-man fire teams, they leapfrog forward, one firing to cover the other. The enemy position is a mess: bunkers

---

 51   UNITED KINGDOM

CVR(T) Scimitar and Scorpion recce vehicles

The CVR(T) – Combat Vehicle Reconnaissance (Tracked) – is the strange designation of a series of light armoured vehicles. Used to scout ahead of the battle group, the CVR(T)s present the enemy with a small, fast moving target, but their armour will only just keep out machine-gun fire. To engage enemy recce forces, Scimitars carry a 30-mm RARDEN cannon; Scorpions are armed with a 76-mm gun. In an infantry battle group, the reconnaissance platoons equipped with them do not hesitate to dismount for a close recce of the enemy's positions at night. There is much more to

With almost no armour protection, the Scimitars and Scorpions rely on their small size and high speed to survive.

reconnaissance to just whizzing around the battlefield, although the Jaguar engines of the CVR(T)s give them sports car acceleration and speed cross-country. These vehicles have proved a popular export and are used by 15 countries other than Britain.

Specification
Crew: 3
Combat weight: 8 tonnes
Engine: Jaguar 4.2-litre petrol engine developing 190 bhp at 4,750 rpm

Top road speed: 80 km/h +
Armament: 1×76-mm gun (Scorpion) or 1×30-mm RARDEN (Scimitar), plus 1×7.62-mm machine-gun

---

 52   UNITED KINGDOM

Sultan command vehicle

The CVR(T) is part of a large family of vehicles based on the same lightweight aluminium-armoured hull. Many others are used within the battle group. The Sultan is designed to provide officers with an armoured command platform, with a troop compartment like an APC, but with a desk along the side, map boards and at least two radios.
Although the back can be extended by fitting a tent around it, the space in a Sultan is very limited and only a small command team can use it successfully. It is used by the commander of a battle group's four self-propelled mortars and for a variety of other specialist roles, including electronic warfare.
The vehicle is very difficult to work

The Sultan command vehicle is very cramped on the move and causes severe motion sickness. Here it is halted with the 'penthouse' tent rigged to give the command team enough space to work in.

from unless the tent can be extended to give more room. A typical command team consists of two officers working on the different radio nets and two soldiers working as watchkeepers. Staying closed down during an exercise simulating the use of NBC (Nuclear, Chemical, Biological) weapons is a severe test of patience and stamina.

Specification
Crew: driver, commander/radio operator plus three or four others
Combat weight: 8.6 tonnes

Engine: Jaguar 4.2-litre petrol engine developing 190 bhp at 4,750 rpm
Top road speed: 80 km/h +
Armament: 1×7.62-mm machine-gun

Left: A Challenger races forward, its 120-mm gun automatically stabilised so that it remains on target even as the tank lurches over the ground.

Below: Despite its enormous size, there is little room to spare inside a tank. Three men are crammed in the turret around the breech of the gun. The driver is isolated in his own compartment in the front of the hull.

53

UNITED KINGDOM 🇬🇧

Chieftain AVLB (Armoured Vehicle Launched Bridge)

There are many different specialist vehicles used by the engineers attached to each battle group. One of the most important is the Chieftain AVLB – an old Chieftain tank with its turret removed and carrying a No. 8 Tank Bridge instead. This arcs high into the air before folding over to bridge a gap just over 22 metres across. This can be positioned very quickly, although the vehicle is easily spotted as its bridge passes through the vertical position during deployment. Many AVLBs carry a track width mine plough on the front to make them dual-purpose engineer vehicles, able to clear a route through an enemy minefield.

An AVLB drops its bridge on the ground during an exercise. Crossing a minor river or anti-tank ditch is a vulnerable moment for the battle group and it must be done quickly.

Specification
Crew: 3
Combat weight: 55 tonnes
Engine: Leyland two-stroke, six-cylinder multi-fuel developing 750 bhp at 2,100 rpm
Top road speed: 48 km/h
Armament: none

54

UNITED KINGDOM 🇬🇧

FV432 mortar carrier

The battle group's artillery support comes from distant batteries of guns that are ultimately controlled by the higher command. The battle group commander does not have the final decision on whether the guns support him or another formation, but he does have a powerful supporting force of his own. Each battle group has four to six 81-mm mortars carried in modified FV432 APCs. With 160 81-mm mortar bombs each, they can fire to a maximum range of 5,600 metres and offer the commander an instantly available source of heavy firepower. They can bombard enemy positions to soften them up, or pin down enemy infantry, or fire smoke rounds to mask

The battle group has its own artillery with it in the shape of 81-mm mortars mounted in FV432s. The mortars are on a turntable with 360-degree traverse so that they can fire in any direction.

the movement of the battle group's own tanks or infantry.

Specification
Crew: 6
Combat weight: 16.4 tonnes
Engine: Rolls-Royce two-stroke, six-cylinder multi-fuel developing 240 bhp at 3,750 rpm

Top road speed: 52 km/h
Armament: 1×81-mm mortar on 360-degree turntable

Target! A direct hit against armour plate gives a blinding white flash and loud explosion.

are caved in by artillery fire and one fire trench has collapsed under the weight of a tank. But no position can be considered taken until infantry have gone firm on top of it.

Commanders' nightmare

The infantry fight through the objective, shooting anything that moves. It's a nightmare for the section commanders, trying to answer calls on the radio while directing the fire of their own men and watching that everybody stays together but without bunching up or wandering into friendly fire. Scrambling over the broken ground to a deafening accompaniment of machine-gun fire and exploding

grenades, the final assault seems to go on for ever. But most soldiers find they have only had time to fire off a single magazine before it is all over.

Only a handful of tanks joined the infantry for the assault. The rest stormed right over the enemy just before the APCs stopped to dismount their infantry. There is nothing like 65 tons of angry tank to keep enemy troops hiding in their bunkers. Now, on the far side of the objective, the tanks form a 'ring of steel', waiting in cover to stop any enemy counterattack dead in its tracks. The infantry secure the objective and reorganise, clambering back into their APCs. The tanks resume their advance and the armoured spearhead of the British Army is on the move again.

A column of T-62s in the advance. An old tank now and no longer found in front-line Russian forces, it is still found all over the world, especially in the Middle East.

Combat Comparison

Two of the most important tanks on the battlefield, the Challenger and T-72, provide a dramatic comparison between the British and former Soviet armies. Both tanks were replaced or updated by more advanced versions: T-72 by the T-80, and Challenger by Challenger II – the same hull but with vastly improved turret with state-of-the-art electronics and a new high-pressure gun.

Challenger

Introduced in 1982 although based on a much earlier design, the Challenger reflects the British thinking on tank design – heavily armoured, with a large rifled gun, and relatively slow. At over £1½m each, the Challenger is much more expensive and complicated than the T-72, and much harder to operate.

The 125-mm smooth-bore gun can fire only two types of round, as opposed to Challenger's three. Ranges are also less.

 **T-72**

The T-72 is widely exported by Russia and is in service in 14 countries. It is much smaller and faster than Challenger, although much more poorly armoured. Designed for use by conscripts, it is simple to maintain.

The Chobham armour of Challenger is vastly superior to that of the T-72. It is almost impenetrable by Soviet-era rounds.

Challenger's crew of four means extra height in the turret. T-72 presents a small and hard-to-hit target.

The 1,200 bhp, 12-cylinder engine drives the Challenger at 56 km/h; the T-72 is less powerful, but faster.

The 120-mm gun is more accurate than that of the T-72, and with computerised fire control has a 97 per cent hit probability.

At 62 tonnes and over 11½ metres long, Challenger is a fuel guzzler. Its range is 402km, but is less than that of the T-72.

With no autoleader to fill the turret, the Challenger can store up to 64 main armament rounds and 4,000 co-axial rounds.

A pack change, although removable in one unit, is still a long and difficult task. T-72 is mechanically simpler.

The gun can only go 5 degrees below the horizontal, so it must expose more turret to shoot downhill.

As with all Soviet tanks, T-72 is designed to deep-ford, unlike any British tank. It takes 20 minutes to prepare.

PETER SARSON / TONY BRYAN

Both tanks are designed to operate in a chemical war. They are fitted with filters, so the crew can fight without masks.

Developments under the Soviet Army meant better frontal armour for the T-72, but it is still woefully poor.

The T-72 has a 12.7-mm commander's gun for anti-aircraft use. Challenger's 7.62-mm weapon has a lower range and less stopping power.

Formation of the BATTLE GROUP

Highly mobile and hard-hitting, the battle group is created for a specific task and given the firepower to carry it out. It's made up of units from every branch of the army; so how do they train to fight together?

Any war in Europe would be dominated by armoured forces: only they have the fire and mobility for a decisive action. But massive tank battles went out of style after the 1973 Arab-Israeli war, when tank attacks were ground down by small groups of soldiers armed with accurate anti-tank guided weapons. So the battle group concept was developed, with all the arms and support units necessary to move, fight and win on the modern battlefield.

Too large a unit and the concentration of vehicles and men becomes an attractive target for tactical nuclear weapons, or a chemical or biological strike. Too small and it will simply be steam-rollered by the enemy armour. It needs to be able to switch rapidly from the assault to defence, and back again, as the need arises. It relies heavily on its own reconnaissance units, both light armoured vehicles and helicopters. Its main striking force is its main battle tanks, but they are backed up by mobile artillery, combat engineers, mechanised infantry and powerful anti-tank units.

Forward reconnaissance

The long arm of the artillery

Most of the artillery support for the battle group comes from a battery of AS 90 155-mm self-propelled guns. The battle group has a Forward Observation Officer permanently attached, whose task it is to control that fire support. If necessary, he can also call on the heavier guns at Division or even Corps level under the British Army's Battlefield Artillery Target Engagement System (BATES).

Left: Until the mid 1980s, the 105-mm Abbott was the standard artillery support weapon for the battle group. It could fire 16kg high-explosive shells out to 17,000 metres.

Below: Reconnaissance means finding out as much as you can about the enemy. Gazelle helicopters are fast, and can scout over a much larger territory than anything on the ground.

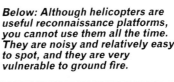

Left: Helicopters are brigade reconnaissance assets, but they can work closely with the battle group's own light armoured reconnaissance platoon.

Below: Although helicopters are useful reconnaissance platforms, you cannot use them all the time. They are noisy and relatively easy to spot, and they are very vulnerable to ground fire.

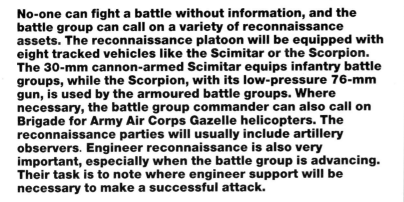

No-one can fight a battle without information, and the battle group can call on a variety of reconnaissance assets. The reconnaissance platoon will be equipped with eight tracked vehicles like the Scimitar or the Scorpion. The 30-mm cannon-armed Scimitar equips infantry battle groups, while the Scorpion, with its low-pressure 76-mm gun, is used by the armoured battle groups. Where necessary, the battle group commander can also call on Brigade for Army Air Corps Gazelle helicopters. The reconnaissance parties will usually include artillery observers. Engineer reconnaissance is also very important, especially when the battle group is advancing. Their task is to note where engineer support will be necessary to make a successful attack.

e Battle Group

LAND ROVER
The long-wheelbase Land Rover is the British Army's standard general purpose light vehicle. Normal load is 850 kg, or eight passengers. It has been made in innumerable versions and can be found from depots in Britain to the front line in Germany.

SCIMITAR Tracked Reconnaissance Vehicle
Small, fast and agile, the Scimitar is the cavalry mount of the 1990s. Traditionally, cavalry was used to scout ahead of the army, seeking out the enemy and observing his movements. It is no different today. The Scimitar ranges ahead of the Battle Group, bringing back the intelligence of enemy positions and movements that any commander needs when making his battle plans.

SPARTAN MILAN Anti-Armour Vehicle
Each British mechanised infantry battalion in Germany has four Spartan APCs fitted with the MILAN Compact Turret. The Spartan carries two missiles ready to launch, with another 11 in the vehicle.

Fighting Vehicles of th

For decades, mechanised warfare has meant tank warfare. At its simplest, during World War II, the Red Army managed to defeat the Germans in the east with huge numbers of just two types of vehicle, the T-34 tank and the American-supplied 2½-ton truck. But warfare has become infinitely more complex since 1945. Tanks are still the mailed fist of modern armies, and trucks are still required to bring up supplies from the rear. However, the highly mobile armies of the 1990s have many specialised battlefield requirements where a tank or a truck just will not fill the bill. So a completely mechanised formation like the Battle Group needs a wide variety of vehicles to carry out all its necessary tasks.

ABBOTT Self-Propelled Gun
Abbott provides the Battle Group's artillery support, but its 105-mm gun is too small for the modern battlefield. It will be replaced in the 1990s by the much more capable AS90 155-mm gun.

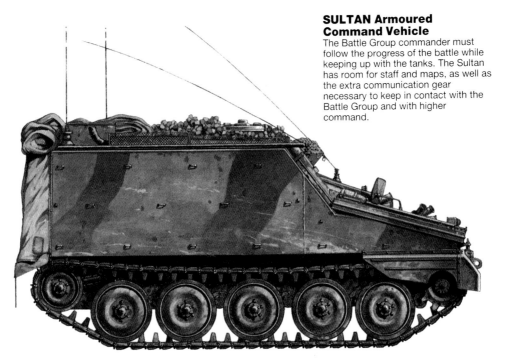

SULTAN Armoured Command Vehicle
The Battle Group commander must follow the progress of the battle while keeping up with the tanks. The Sultan has room for staff and maps, as well as the extra communication gear necessary to keep in contact with the Battle Group and with higher command.

Iron fist of armour

The power of the battle group lies in its tanks. With an unmatched combination of firepower, protection and mobility, the main battle tank has ruled the battlefield since World War II, and remains vital even in this age of the guided missile. The battle group is a mixed armour-infantry formation of battalion size, in which the tanks provide punch. The proportion of tanks to infantry depends upon the task in hand, but in an infantry battle group there will be one or two 14-tank squadrons serving alongside two or three infantry companies. In an armoured battle group the proportions are reversed, with two or three tank squadrons being matched with one or two infantry companies.

Left: A Challenger main battle tank sits in an ambush position, behind an earth bank. Both armoured and infantry battle groups field tanks in 14-tank squadrons, the armoured group being tank heavy and the infantry group having more infantry.

When it was first introduced in the 1960s, the Chieftain's 120-mm gun and heavy armour made it the most powerful tank in the world. Although it has since been outclassed by more modern designs, it remains an extremely potent fighting machine.

Below: In the UK, the Chieftain was replaced by the Challenger (illustrated). This is a much more modern design, which retains the 120-mm gun of the older tank but which is even better protected.

Above: A Challenger main battle tank roars into action. Post-war British tanks have tended to be heavily armed and armoured vehicles, and the Challenger is no exception. But it has a much more powerful engine, a better suspension, and a better transmission system than the preceding Chieftain, which gives it superior cross country performance. This means that Challenger-equipped battle groups are very fast-moving formations.

Infantry assault

Left: A MILAN vehicle crosses an AVLB-laid bridge. The compact turret weighs under 200 kilos, and is fitted on the Alvis Spartan armoured personnel carrier. Using the vehicle gives the MILAN teams the mobility to operate with the tanks.

Below: Infantry burst out of their Warrior infantry combat vehicle. In an attack, the infantry is carried right up to the enemy lines, supported by fire from tanks that have burst right through the enemy positions and by tanks on the flank.

While main battle tanks are superb offensive or defensive fighting machines, they cannot take and hold ground without a considerable amount of preparation, and they lose their essential mobility. Tanks are also vulnerable to well-sited anti-armour weapons. To protect the armour from such threats, and to ensure that any ground taken is retained, you need infantry. But the infantryman is not the footslogger he would have been a generation ago. In modern warfare the mechanised infantryman is carried right up to the enemy lines in an armoured personnel carrier or an infantry fighting vehicle. This is usually armed with a cannon so as to provide close range fire-support when the infantry dismount to make the final close-range assault on foot.

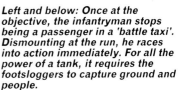

Left and below: Once at the objective, the infantryman stops being a passenger in a 'battle taxi'. Dismounting at the run, he races into action immediately. For all the power of a tank, it requires the footsloggers to capture ground and people.

Above: MILAN is a German/French anti-tank missile made under licence in the UK. The infantry version is manned by a crew of two, but in spite of being called 'man portable' the system is quite a load to lug around.

Left: MILAN provides the bulk of the battle group's anti-tank guided weapon capability, especially when mounted on the MCT Spartan APC. The MILAN Compact Turret mounts a pair of ready-to-launch missiles, with a further 11 missiles inside the vehicle. The missiles are identical to the infantry version and can be fired from a standard MILAN firing post.

ARMOURED INFERNO

EYE WITNESS

"The German commander had not intended to commit his Panthers quite so soon, but once they had made sudden contact with the advancing British armour it proved impossible to disengage. The remainder of the afternoon and evening turned into the biggest tank battle of the Normandy campaign, a shootout between the outnumbered but superior Tigers and Panthers and our massed but more vulnerable Shermans and Cromwells".

General Richard O'Connor, Commander, VIII Corps, British Army

The World War II British armoured division, precursor to the modern battle group, was a combined-arms unit of considerable power. Its test came in Normandy in 1944, where it met the battle groups of the German Panzer Divisions.

The German soldiers east of Caen were weary. They had been fighting since the massive Allied invasion of Normandy six weeks before. They had managed to contain, at fearful cost, the overwhelming power of the British and Americans to a strip along the coast.

But 18 June was different. On both sides of the line, armies were woken by a new sound – a droning hum, coming from the north, swelling gradually to a mind-numbing roar. Overhead, one thousand heavy British bombers filled the sky. They were joined by over a thousand heavy and medium bombers of the 8th and 9th US Army Air Forces, in the largest single bombing raid of its type in history.

British soldiers watched in wonder as this vast armada passed overhead, but the

Above: The body of a German infantryman and a silent Panther tank are testimony to Allied power in the air over Normandy. The massive bombardment before Operation Goodwood was expected to leave all the German defences in this state, but British tank crews were in for a nasty shock.

Right: During Operation Goodwood, the infantry was left far behind the charging armour, leaving the British tanks vulnerable to German counter-attack.

Right: Cromwell tanks form up before crossing the River Orne. With only six bridges available, each of the three armoured divisions involved in Operation Goodwood had just two bridges across which to move more than 3,000 vehicles. The monumental traffic jam meant that many tanks never even got to the battle.

Left: Bombs dropped from a flight of Douglas A-20 Havoc bombers of the US 9th Air Force explode in the Norman countryside. More than 2,000 medium and heavy bombers dropped thousands of tons of high explosive on German positions as Bomber Command and the US Army Air Force made a maximum effort in support of Montgomery's Goodwood offensive.

Above: Knocked out tanks can often be salvaged, so recovery tanks are an important part of any armoured force's equipment. Here a Sherman recovery tank pulls a crippled gun tank through Bourguébus.

Above: German soldiers on the invasion front move carefully through a village. The defenders of Normandy were a mixed bag; garrison divisions of dubious fighting quality served alongside SS and Panzer divisions hardened by years of combat in Russia.

Germans in the narrow corridor east of Caen felt only terror when the bombs started dropping. Soon a storm of high explosive was reshaping the Norman landscape. "It was a bomb carpet," one German soldier was to write, "regularly ploughing up the ground. Amidst the thunder, we could hear the screaming of the wounded and the insane howling of men driven mad."

Even as the bombers departed, they were followed by the crash of a rolling barrage from two hundred British artillery pieces, ranging from 25-pounders to the massive 15-inch guns of the monitor HMS *Roberts* off the coast. And under the cover of that barrage, nearly 900 tanks from three British armoured divisions were moving forward. Operation Goodwood was under way.

Moving slowly

The Allied invasion of Normandy had bogged down. Targets which had been scheduled for capture in the first few days of the invasion were, thanks to stubborn German resistance, only now being approached. But now General Sir Bernard L. Montgomery had 13 divisions of his 21st Army Group ashore, fully supplied and ready for action. He had evolved a strategy to involve the Germans on the east of the brid ehead in a grinding battle of attrition, drawing in their reserves and leaving an opening for the US First and Third Armies to break out in the west. Operation Goodwood was Montgomery's opportunity for "a real show-down on the eastern flank, and to release a corps of three armoured divisions into the country about the Caen-Falaise road."

General Sir Richard O'Connor's VIII Corps, made up from the 7th, 11th, and Guards Armoured Divisions, had begun crossing the bridges over the River Orne during the night of 17 July. Following the massive bombardment, the divisions began their movement forward, tanks in the lead, followed by the infantry

in their half-tracks and tracked carriers (often known as 'Bren-gun carriers').

The immediate opposition had been shattered by the rain of high-explosive, and the 16th Luftwaffe Field Division had been all but wiped out. But the commander of one of the battlegroups of the 21st Panzer Division was very much alive. As the first British tanks pressed into the corridor between Caen and the Bois de Bavent, Oberst Hans von Luck was returning to the battlefield from three days' leave in Paris.

Scraping together what defences he could, most notably some of the deadly Luftwaffe 88-mm Flak guns, he engaged tanks of the 11th Armoured Division. Twelve tanks were smashed within minutes. However, several squadrons had already pushed on southwards before von Luck's arrival. The German colonel went on to round up what he could from the isolated pockets of armour which had survived the bombardment.

By the middle of the morning, 11th Armoured was approaching Bourguébus ridge, which was commanded by a battery of German 105-mm self-propelled guns. In spite of losing some tanks, it was within sight of its first objective. The Guards Armoured Division was coming up behind, although it too was losing some tanks to von Luck's battle group. 7th Armoured was still crossing the Orne far to the rear.

The operation appeared to be going pretty much according to plan, except that the tanks had outrun its halftrack-borne infantry. And they were going beyond the range of the artillery, which was still stuck on the other side of the Orne. Still, with luck the three divisions might manage a breakout into the French countryside. It hadn't been planned, but if the opportunity presented itself . . .

Unfortunately, it was now the Germans who acted. Warned of the coming attack by reconnaissance the night before, the Germans had moved part of the 21st Panzer Division back from the corridor, and had moved the 1st SS Panzer Division from its positions south of Caen. The *Leibstandarte Adolf Hitler* had fought everywhere from Poland to Italy, and it was manned by some of the toughest soldiers and tankmen in the world.

From positions on Bourguébus ridge, the German combat veterans could look down on the British tanks as they came towards the ridge. The German defences were thorough. Farmhouses, coppices, hedgerows; all sheltered anti-tank guns or tanks including a number of the powerful Tigers. Each strong point was protected by infantry and machine-gun positions. And the British attack was heading for this defence in depth along a corridor of open ground as little as three miles wide.

As the tanks of the Fife and Forfar Yeomanry began coming up the ridge, "the fun began," as one British tankman put it.

"I saw Sherman after Sherman go up in flames, and it got to such a pitch that I thought that in a few minutes there would be nothing left of the regiment." 11th Armoured's Sherman and Cromwell tanks were no match for the Panthers of *Leibstandarte*.

The 23rd Hussars, following up, also made an attempt on the ridge, but the wrecked Shermans lying about together with fire from

Above and right: Artillery was one of the few areas where the best of German equipment did not outperform that of the Allies. But when the tanks of the 11th Armoured Division went beyond the range of their own artillery, that didn't matter. A battery of German 10.5-cm howitzers (above) mounted on French Lorraine chassis was in place on Bourguébus ridge when the British attacked, and did considerable damage to the Cromwells and Shermans coming up the slope. Had British weapons like the 5.5-inch Medium Gun (right) been close enough to give the tanks the support they needed, things could have been different.

PzKpfw V Panther

Designed to counter the Soviet T-34, the Panther was to become the best fighting vehicle of the war. Allied tanks in Normandy were no match for the Panther individually, and it normally took five of them to beat one of the German tanks.

Performance
The Panther was powered by a water-cooled V-12 petrol engine developing 700 bhp and which was capable of driving the 44-tonne vehicle at 45 km/h (30 mph)

Crew
Panther carried a crew of five, with a driver and machine-gunner/radio operator in the hull and the commander, gunner, and machine-gunner/loader in the turret.

The Professional's View:

The Panther

"The Panther was the best tank of the War. In Normandy, we could stand off beyond the range of the British and American tanks and could shoot them without any reply. Of course, we had to be alert, because if they got close enough they could still do us some damage.

Panther gunner

Gun
Panther's long 75-mm gun could fire high-velocity rounds capable of penetrating 120mm of sloped armour at 1000 metres.

Armour
Panther's armour ranged from 20-mm of plate under the belly to over 120mm on the well-sloped front of the hull and on the turret.

the Panthers convinced them to withdraw. Even as they did so, however, fire from a village to one side hit and set ablaze every tank in the Regiment's 'C' squadron.

Behind the 11th, the Guards Armoured Division had been engaged by the German troops by-passed in the first stages of the armoured charge. These included several Tigers together with well positioned anti-tank guns in a patch of woods near Cagny, which had miraculously escaped the bombing at dawn. The third element in the attack, the 7th Armoured Division, had hardly got over the river, such was the congestion at the Orne bridges.

The net result of Goodwood was that the British had taken some six miles of ground, but at the cost of 270 of the 870 tanks engaged. Human casualties were surprisingly light for such a fierce fight. The tank regiments of the 11th Armoured Division, which had borne the brunt of the fighting, lost about 80 men, while their supporting infantry, which had followed the tanks all day in their carriers and half tracks, lost only 20.

Traffic jam

In tactical terms, Goodwood was not a success. The massive bombardment had indeed disrupted the German defences, but, with that flexibility that hard years of warfare on the Eastern Front had taught, they had formed strongpoints and ad hoc battle groups on the bypassed terrain.

Of course, the attack was not helped by starting out with three entire armoured divisions, some 9,000 vehicles, trying to cross just six bridges on the Orne. This created a monumental traffic jam, which meant that 7th Armoured, the third to cross, hardly had a chance to get into the action. The area of battle east of Caen was too small, so there was no real room for large formations to manoeuvre.

The charge of the 11th Armoured Division's tanks was in the finest traditions of the cavalry of old, but it meant that they outran their infantry and were soon beyond the range of their own supporting artillery. This proved costly when the division encountered a network of anti-

tank guns and artillery deployed in depth, each position protected by infantry and machine-guns, and supported by small but significant numbers of powerful Panther and Tiger tanks.

Although little ground was taken during Goodwood, it was not entirely a failure. British losses of equipment could be made good in days, but German tank losses were not so easily replaced. And Montgomery's strategic aim was still valid. Such a powerful attack sucked in much of the German reserve armour, and repeated attacks on other sectors around Caen in the following days meant that the Panzer divisions could not withdraw. Thus they were in no position to challenge General

Germans surrender to the British in Normandy. The tree-lined, sunken lanes so typical of the Norman bocage country are ideal for defenders but murder for attackers, as each copse and farmhouse could conceal an anti-tank position.

George S. Patton when the American Third Army broke out from Normandy a month later.

Goodwood lessons are visible today. Tanks on their own suffer when faced with anti-armour defences. The basic unit of the modern British armoured division ensures that the tanks are protected. The Battle Group is a combined arms formation in which tanks provide the punch but operate in close association with infantry, artillery, and anti-tank weapons.

Cromwell Cruiser Tank

The Cromwell was a classic British Cruiser tank: fast, but lightly armed and armoured. Cruiser tanks were Cavalry weapons, for use in high-speed pursuits, but less effective in battle with other tanks.

Gun
Originally armed with the totally inadequate 6-pounder gun, the Cromwell was fitted with a more powerful 75-mm weapon. This could fire high-explosive as well as anti-tank rounds, but while much better than the 6-pounder it could not match the range or penetration of the high-velocity German 75-mm gun fitted to the Panther.

Crew
Cromwell had a five-man crew, with two in the hull and three (commander, gunner and loader) in the turret.

Armour
Although more heavily protected than previous cruisers, the Cromwell's armour, with a maximum thickness of 102 mm, was no match for the Panther's gun.

Performance
Cromwell was powered by a Rolls-Royce V-12 water-cooled engine, which could have given the tank a top speed of 64 km/h. Unfortunately, the Cromwell's suspension could not handle that, so speed was governed to a 52 km/h maximum.

Ring of STEEL

With over 500 men and 200 vehicles to control, the battle group attack is one of the hardest tactical operations there is. It needs accurate information, well-rehearsed drills and precision timing. But when it goes in, the effects are devastating.

1 Timely and accurate information is the key. Your recce platoon must see without being seen, and observe without giving themselves away. The plan can only be as good as the information they give you, so the very best soldiers should be in the recce platoon.

2 Swamp the enemy, and use maximum shock action. He must not know what has hit him until too late. You should try to achieve a 3:1 superiority, so if there are five tanks on the enemy position you will need at least 15.

3 Know your enemy. Put yourself in his shoes. If you attack the enemy from the direction he wants, you will drive staight into a trap. Much better to surprise him by coming from a different approach.

4 Expect the unexpected. Always be on the look-out for the counter-attack. An enemy force may be held in reserve to retake the position if you take it.

5 Artillery is a battle-winning weapon. In the battle group you will only have one battery (eight guns) at your call. You may need more. Your artillery officer (the Battery Commander, or BC) will request them for you.

6 Remember your engineers. Don't be forced by the enemy's obstacle plan to go in the direction he wants you to go: your

engineers can breach his obstacles. Don't forget that an obstacle is only an obstacle if it is covered by fire, so your engineers will need local protection.

7 At the end of the day the only people who can clear the enemy are the PBI – the poor bloody infantry. They will be in the trenches face to face. The bayonet is a weapon that then comes into its own!

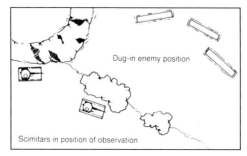

Dug-in enemy position

Scimitars in position of observation

1 Recce ahead

Always travelling in front of the battle group will be the Scimitars of the recce platoon. Their task is to probe the enemy and discover as much information as possible for the Battle Group commander. They will also pass back information on routes, obstacles and minefields. Sometimes working with Gazelle recce helicopters, which can zoom across the battlefield to investigate contacts, they are the 'eyes and ears' of the battle group.

This is the final assault of an enemy position by a combined all-arms battle group – the climax of the battle group attack. The well-thought-out plan, tightly controlled and efficiently executed, brings all the different parts of the team together. But when it comes to the assault, only the tanks and the infantry can get in with whatever it takes.

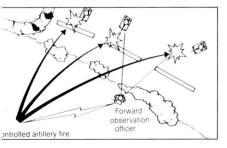

2 Command

With the up-to-date information from his recce platoon, the commander can make his plan. He will issue his orders by radio as soon as possible. An artillery Forward Observation Officer will bring in an accurate and destructive fireplan on the enemy to cover the battle group as it moves into attack formation.

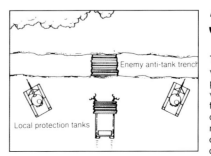

3 Obstacle breach

The enemy has sited his position well and has included an obstacle plan. But the armoured engineers will go forward, protected by the tanks, and lay fascines to fill the anti-tank ditch. They may also be required to breach minefields, clear wire, lay bridges or perform any of a hundred other tasks.

TACTICS

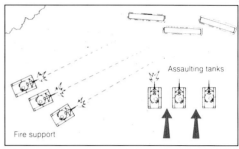

4 Fire support

Once the orders have been given, the battle group will move with well-rehearsed precision. Some of the tanks will move off to a flank, where they will provide the fire support to the attack. Their task is to kill any armour on the enemy position and to keep the enemy infantry from firing back. They will open fire just before the attack is launched for maximum surprise. Once the attack is under way, they will switch to protect the attacking tanks from any possible surprise counter-attack coming in from a flank.

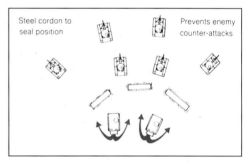

5 Ring of steel

At H-hour the attack will be sprung. The artillery will be firing onto the enemy, including the mortars. The tanks from the fire support will be shooting in the assault. Some will drive around the enemy position, shooting into it, but making sure they keep out of range of any hand-held anti-tank weapons on the position. Once they are past the enemy they take up fire positions to form a 'ring of steel' to cut off the position and prevent any reinforcement or counter-attack.

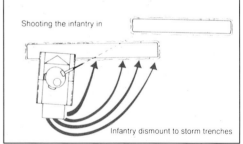

6 Infantry assault

Following behind the assault tanks will be the infantry in their APCs; they will be guided onto the position and supported by a troop of tanks known as 'intimate support tanks'. These tanks provide the platoon commander with massive overwhelming local firepower at his immediate command. The APCs will stop at the last possible moment and the infantry will burst out to storm the trenches. This is the real hand-to-hand combat, which requires guts and determination as well as courage and aggression. It is now up to them to clear the trenches and kill the enemy.

RECCE VEHICLES

Reconnaissance wins battles. Finding out about the enemy is the name of the game: a game that hasn't changed for centuries. Horsemen used to do the job, but modern reconnaissance units have a variety of powerful vehicles to choose from.

Charging around in the high-speed armoured dune buggy or FAV of the US Marine Corps with its 40-mm grenade launcher and other equipment may be a heck of a lot more fun than sitting in the snow in a greatcoat somewhere in Russia, but you can be sure that the men in both situations are equally well trained and professional.

Concealed by the early morning mists of 13 May 1940, the reconnaissance battalion of Major-General Erwin Rommel's 7th Panzer Division stealthily approached the French defences on the far bank of the River Meuse. Their role was not only to observe the enemy positions but, if the opportunity arose, to fight their way across the river. Noticing how unprepared the French were, the recce battalion CO ordered an immediate attack.

The motorcycle teams cocked their MG 34 machine-guns while the SdKfz 231 armoured cars moved into position to provide fire sup-

port. Suddenly the air erupted with the noise of rifle fire as the German assault companies went in. Then came the rattle of machine-guns and the crack of the 20-mm guns of the SdKfz 231s. Totally surprised by the attack, the French were too stunned to react, allowing the assault troops to gain a foothold across the river.

Textbook example

Meanwhile, main force units from the Panzer division had raced forward to support the reconnaissance troops. Within a few hours the Germans were racing forward over the Meuse, the French defences fatally breached. This was a textbook example of reconnaissance by force.

Before World War I, reconnaissance was the function of the cavalry arm, but the advent of machine-guns and the accurate long-range artillery, combined with motor vehicles, transformed this vital aspect of warfare. Even at the start of World War I, Rolls-Royces were being converted into armoured recce vehicles. In the inter-war period, specialised or dedicated reconnaissance vehicles came into their own, and were used to good effect in conflicts such as the Russian and Spanish Civil Wars. During World War II the reconnaissance vehicle achieved a new importance, especially in the mechanised columns that

made Blitzkrieg warfare possible.

Since 1945, the armoured and scout cars of that war have been supplemented by vehicles ranging in size from jeeps to Main Battle Tanks. The reason for this extraordinary diversity of types is in part because different forms of reconnaissance require different vehicles, but also it reflects varying

A British Scimitar and a US HMMWV (pronounced Hum-vee) of the ACE Mobile Force are seen on exercise in Norway. The AMF is NATO's strategic reserve, kept under direct command of the Supreme commander and ready for emergency deployment. It is made up of troops from all NATO countries.

approaches to the gaining of information. Two rival schools of thought have developed: they can be summarised as 'recce by stealth' and 'recce by force'.

Recce by stealth

The British and Commonwealth armies are the foremost representatives of the 'recce by stealth' approach, where reconnaissance is kept as low-key as possible. Information is gained by concealment and fieldcraft, so that recce troops act as scouts for the main body. Consequently their vehicles are lighter and less well armed than those of forces that employ 'recce by force' tactics. Thus, for example, while the British Scorpion CVR(T) weighs just over 8 tonnes and is armed with a low-velocity 75-mm gun, the French AMX-10RC is nearly twice as heavy and is

FLASHBACK

The Royal Northamptonshire Yeomanry are shown on patrol on the Western Front in 1914. Having deployed on horse, the patrol dismounts to do its probing recce on foot.

Horses on patrol

The execution of battlefield reconnaissance had always been the realm of the light cavalry. The light horse was their vehicle; swords or, later, a rifle their firepower (the heavier horses of the Dragoons and Dragoon Guards were the armour of those days). Often mounted in small patrols of a few horses, the scouts would infiltrate well forward of their lines and, using terrestrial telescopes, would acquire as much information as possible before riding like fury back to their own lines to report. The evolution of the machine-gun and radio put paid to horse recce, although the idea of a small, highly agile, lightly protected recce beast remains.

RECCE VEHICLES Reference File

99
GERMANY

Thyssen Henschel Spähpanzer 2 Luchs

After the re-formed German army had gained initial experience it ordered the development of 4×4, 6×6 and 8×8 vehicles as the basis of a complete family of vehicles. One of the 8×8 vehicles was the **Spähpanzer 2** designed by Daimler-Benz. Trials were completed in 1971, and deliveries were made between 1975 and 1978.

The vehicle is named **Luchs** (lynx) in service. An unusual feature is the location of the radio operator at the left rear of the vehicle, facing backwards, so that he can double as a second driver. The two-man powered turret contains a dual-feed cannon, and this allows immediate switch from HE to armour-piercing ammunition if a light armoured fighting vehicle is spotted.

Specification
Spähpanzer 2 Luchs
Type: four-crew wheeled combat reconnaissance vehicle
Weight: 19.5 tonnes (19.19 tons)
Armament: one 20-mm cannon with 375 rounds, one 7.62-mm (0.3-in) co-axial machine-gun with 100 ready-use rounds.
Performance: maximum speed 90 km/h (56 mph); range 800 km (497 miles)
Dimensions:
length overall 7.743 m (25 ft 4.8 in); width 2.98 m (9 ft 9.4 in)
Powerplant: one 291-kW (390-hp) Daimler-Benz OM 403A multi-fuel engine
User: Germany

A Bradley M3 Cavalry Fighting Vehicle gets some welly. Now the standard recce vehicle of the US armoured division, the Bradley is a little on the large size for its role. It is made entirely out of aluminium in an attempt to keep its weight down, but this means that it has questionable armoured protection.

Left: The South African Eland with 90-mm gun fights a night battle. Based on the French-designed AML, it proved its worth in the Namibian and Angolan campaigns against Soviet-made T-34s.

The Professional's View:

Recce squadron

"I got the impression that the men in recce had been posted for some sort of 'individualism' that had proved unwelcome in the staid ranks of the tank squadrons. At first, I observed that my comrades had been selected for a variety of qualifications which seemed somewhat dubious to me. On arrival in Normandy I was soon to discover that the squadron included some of the finest NCOs and men I could wish to command. Exactly that individualism and resourcefulness proved to be a major asset and, had I tried, I could not have picked a better lot of soldiers. During the entire campaign we had only 10 casualties — testimony to their proficiency and 'individualism'."

Officer Commanding, armoured recce squadron, 1945

armed with a high-velocity 105-mm gun.

The Germans pioneered reconnaissance by force during World War II, and their successes in that conflict greatly influenced other nations, including America, France and the Soviet Union. The advantage of 'recce by force' is that larger and better-armed vehicles allow reconnaissance troops to operate further from their parent body – perhaps as much as 60 miles – and for longer periods. In addition, by attacking hostile outposts, the recce unit forces the enemy to expose his positions and capabilities in the act of defending himself. There are disadvantages to this approach, though: the reconnaissance unit

---

100
FMC Lynx

CANADA 🇨🇦

The West's most widely produced armoured personnel carrier is the M113 series, and when this type entered production in 1960 FMC appreciated that major subassemblies could be used in other armoured vehicles. A good example is the **Lynx Command and Reconnaissance Vehicle**.

The type has the same type of welded aluminium construction as the M113. The commander sits in the centre of the vehicle under a large manually operated turret fitted with an externally mounted heavy machine-gun. The Dutch vehicles were delivered with the same M26 turret, but it was replaced by a 25-mm KBA cannon. The dual-feed cannon is provided with 120 HE and 80 APDS rounds.

Specification
FMC Lynx Command and Reconnaissance Vehicle
Type: three-crew tracked combat reconnaissance vehicle
Weight: 8.775 tonnes (8.64 tons)
Armament: one 12.7-mm (0.5-in) machine-gun with 1,155 rounds, one 7.62-mm (0.3-in) machine-gun with 2,000 rounds.
Performance: maximum speed 71 km/h (44 mph); range 523 km (325 miles)
Dimensions: length overall 4.597 m (15 ft 1 in); width 2.413 m (7 ft 11 in)
Powerplant: one 160-kW (215-hp) Detroit Diesel 6V-53 diesel
Users: Canada and Netherlands

SOVIET TANK DIVISION IN THE ADVANCE

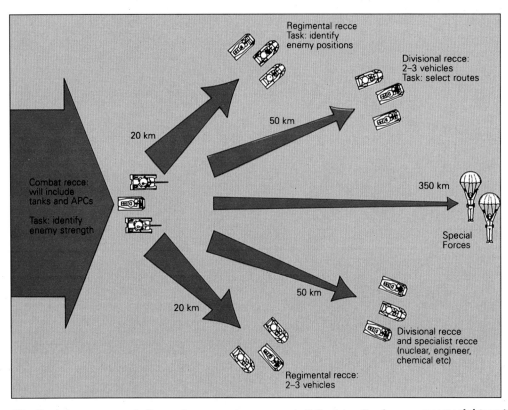

Combat recce: will include tanks and APCs

Task: identify enemy strength

20 km

50 km

Regimental recce
Task: identify enemy positions

Divisional recce: 2–3 vehicles
Task: select routes

350 km

Special Forces

20 km

50 km

Divisional recce and specialist recce (nuclear, engineer, chemical etc)

Regimental recce: 2–3 vehicles

The Soviets were great believers in reconnaissance, and all Soviet units, from company right up to a multi-army front, had specialist recce formations.

can find itself involved in a firefight for which it is not fully equipped, its vehicles are more expensive and, because of their increased weight, are less manoeuvrable.

Soviet tactics

The Soviet Army displayed an interesting and novel attitude in this field, as its reconnaissance units were made up of a mixture of vehicles with very different capabilities. A typical patrol might consist of a T-72/64 MBT that provided the unit with firepower from its powerful 125-mm main gun; one or two BMP-1 Infantry Combat Vehicles; and a BRDM-2 scout car, which although a wheeled vehicle had sufficiently good cross-country mobility to keep up with the other tracked elements.

The US Army, unlike most other armed forces, continued to employ and develop light tanks in a reconnaissance role after the end of World War II. The M41 light tank served throughout the 1950s and was replaced in the following decade by the M551 Sheridan light tank, which was armed with the 152-mm Shillelagh gun/missile system as a means of uprating firepower without the need for a heavy main gun. The Sheridan was not a success, however, performing poorly on active service

---

101

CANADA

General Motors of Canada LAV

The origins of this important vehicle lie with a Swiss vehicle, the MOWAG Piranha series of armoured personnel carriers. The 6×6 version was adopted for Canadian production, and the 8×8 model selected by the US Marine Corps to meet its **Light Armored Vehicle** requirement.

The initial version is the **LAV-25**, which can carry six infantrymen in addition to its three-man crew. The latter include the commander and gunner in the powered Delco turret. This is of the same welded steel construction as the hull, and its fully stabilised main weapon is the M242 Bushmaster, a dual-feed Chain Gun cannon. It has standard features such as an NBC system.

Specification
General Motors of Canada LAV-25
Type: three/nine-crew wheeled multi-role combat vehicle
Weight: 12.882 tonnes (12.68 tons)
Armament: one 25-mm cannon with 630 rounds, one 7.62-mm (0.3-in) co-axial machine-gun and one optional 7.62-mm (0.3-in) AA machine-gun with 1,620 rounds
Performance: maximum speed 100 km/h (62 mph); range 668 km (415 miles)
Dimensions: length overall 6.393 m (20 ft 11.75 in); width 2.499 m (8 ft 2.5 in)
Powerplant: one 205-kW (275-hp) Detroit Diesel 6V-53T diesel
User: USA

102

FRANCE

AMX-10RC

In the late 1960s France started work on a powerful reconnaissance vehicle to replace the Panhard EBR. Considerable saving of time and cost was achieved by use of many features from the AMX-10P infantry fighting vehicle. Trials confirmed that the **AMX-10RC** was little less than a wheeled tank in offensive terms. Production began in 1978 and ended in 1987.

The driver can vary the suspension height of the 6×6 drive to suit the terrain being crossed, and two waterjets are used for amphibious propulsion. The vehicle has NBC protection and passive night vision equipment.

Specification
Atelier de Construction Roanne AMX-10RC
Type: four-crew wheeled combat reconnaissance vehicle
Weight: 15.88 tonnes (15.63 tons)
Armament: one 105-mm (4.13-in) gun with 38 rounds, one 7.62-mm (0.3-in) co-axial machine-gun with 4,000 rounds
Performance: maximum speed 85 km/h (53 mph); range 800 km (497 miles)
Dimensions: length overall 9.15 m (30 ft 0.25 in) with gun forward; width 2.95 m (9 ft 8.1 in)
Powerplant: one 194-kW (260-hp) Renault HS 115 diesel
Users: France and Morocco

in Vietnam.

A major advance for the US has been the development of the Lynx command and reconnaissance vehicle. Based upon the M113A1 armoured personnel carrier, the Lynx has been exported to Canada armed with 12.7-mm and 7.62-mm machine-guns and to Holland, where the heavy machine-gun is replaced by a 25-mm cannon housed in its own special turret.

West Germany has carried on the tradition of producing superb armoured fighting vehicles that it started during World War II. In the field of reconnaissance it has the Spähpanzer 2 Luchs, an 8×8 reconnaissance vehicle that is

Above: The Panhard ERC-90 F1 Lynx is seen in service with the Mexican Army. It is a simpler vehicle and thus much cheaper than the AMX-10RC.

Right: Motor bikes may seem an obvious choice for recce, but they are too vulnerable in anything less than a low-intensity conflict such as Namibia.

103
FV721 Fox

UNITED KINGDOM

The **Fox** was developed in 1965 and introduced into service with the British Army in 1970. It was used in the UK as a reconnaissance vehicle for about 10 years, though today only Malawi and Nigeria keep it in service.

The Fox is a direct development of the Ferret Scout Car and has an all-aluminium hull designed to afford protection against small-arms and shell splinters while keeping weight to a minimum. It is fitted with the 30-mm RARDEN cannon and co-axially mounted 7.62-mm machine-gun. Other variants include the **Panga**, fitted with a 12.7-mm M2 HB machine-gun, and the **Fox 25**, fitted with a 25-mm Hughes M242 Chain Gun.

Specification
FV 721 Combat Vehicle Reconnaissance (Wheeled) Fox
Type: three-man wheeled armoured reconnaissance vehicle
Weight: 6.12 tonnes (6.02 tons)
Armament: one 30-mm (0.12-in) cannon with 99 rounds, one 7.62-mm (0.3-in) co-axial machine-gun with 2,600 rounds, and four smoke-dischargers on each side of the turret
Performance: 104 km/h (64.6 mph); range 434 km (270 miles)
Dimensions: length overall 5.08 m (6 ft 8 in); width 2.134 m (7 ft); height 2.2 m (7 ft 3 in)
Powerplant: one 141-kW (190-bhp) Jaguar XK petrol engine
Users: Malawi, Nigeria, UK

104
BRDM

FORMER USSR

In the late 1950s the USSR introduced the **BRDM-1** scout car. The BRDM-1 is a conventional 4x4 type with common Soviet-era features such as a central tyre pressure-regulation system and a single waterjet for amphibious propulsion. It also has two pairs of semi-retractable belly wheels which can be lowered for additional chain-driven traction under adverse conditions.

In the early 1960s the improved **BRDM-2** was introduced with a more powerful engine, greater fuel capacity, NBC protection, better vision for the commander and driver, and turreted armament in place of the BRDM-1's single pintle-mounted 7.62-mm (0.3-in) machine-gun. The BRDM-2 has also been developed in command,

radiological reconnaissance and anti-tank variants.

Specification
BRDM-2
Type: five-crew wheeled scout car
Weight: 7.0 tonnes (6.89 tons)
Armament: one 14.5-mm (0.57-in) machine-gun with 500 rounds and one 7.62-mm (0.3-in) co-axial machine-gun with 2,000 rounds
Performance: maximum speed 100 km/h (62 mph); range 750 km (466 miles)
Dimensions: length overall 5.75 m (18 ft 1.3 in); width 2.35 m (7 ft 8.5 in)
Powerplant: one 104-kW (140-hp) GAZ-41 petrol engine
Users: 46 countries worldwide

The Ferret scout car is no longer used in the recce role, but it had a long career in that task. Now pushing 40, it remains in service with practically every regiment and battalion in the British Army.

well armoured, fully amphibious and has an exceptional operational range of over 500 miles. Other equipment includes an NBC system and advanced night-vision infra-red equipment.

However, one area the German military has moved away from since 1945 is that of armament. Whereas the recce AFVs of the Wehrmacht carried guns of 50-mm and 75-mm calibre, the Luchs is armed with a high- velocity Rheinmetall Mk 20 Rh 202 cannon. The German concept of reconnaissance by force has obviously been moderated by the extreme demands of warfare in modern theatres.

"Time never wasted"

The French – always with an eye on the export market in Africa – have manufactured their own very particular reconnaissance vehicles. Top of the range is the very expensive 6×6 AMX-10RC. This is supplemented by the Panhard ERC Sagaie armoured car, which, although a very much lighter vehicle, is armed with a 90-mm gun.

The most prolific of French reconnaissance vehicles is the Panhard AML-90, which again is fitted with a 90-mm gun. The AML-90 is built under licence by Sandock Austral in South Africa for its own armed forces and, renamed the Eland, it has been very successful in anti-guerrilla operations. In Africa, heavily armed reconnaissance armoured cars have acquired a new role as major offensive vehicles in their own right, especially for internal security operations – hence the success of Panhard armoured cars.

Looking towards the future, there can be no doubt that reconnaissance vehicles have an important role to play in the battlefield of the future. Some of the most interesting work in this area is being carried on by the US armed

forces. The Americans have been working with light tanks for their Rapid Deployment Force, and have produced a number of vehicles, including the RDF Light Tank, equipped with an automatic magazine. In addition, the American Light Armoured Vehicle (LAV) programme selected the 8x8 General Motors Canada Piranha, which comes in several configurations, including the main vehicle, armed with a 25-mm Chain Gun, an assault version with 90-mm main gun, and a cargo carrier.

Of course, on the battlefield of today and that of the future, land-based reconnaissance is only one of many forms of tactical and battlefield recce, supplemented by information from aircraft, helicopters and pilotless drones (RPVs). Nonetheless, the troops on the ground have an important role to play; and the old military saying still applies: "Time spent on reconnaissance is never wasted."

Not too difficult to spot: the M551 Sheridan light tank fires its 152-mm Shillelagh missile. Now phased out as the US recon vehicle, it was never terribly good.

Recce Rival War II

The Allies tried several tactics for their recce units, and there was no set pattern even at the end of the war. Their vehicles tended to be smaller, lighter and armed only for self-defence.

105

  USA

Light Armoured Car M8

In 1940 and 1941 the US Army was able to observe operational trends in Europe and so develop a new armoured car with a good performance, a 37-mm (1.45-in) gun, 6×6 drive, a low silhouette and light weight. Four companies were ask to tender, and of these Ford produced the **T22**, which was to become the **Light Armoured Car M8**.

The M8 subsequently became one of the most important of all the American armoured cars, and by the time production ceased in April 1945 no fewer than 11,667 had been produced. It is an indication of the vehicle's success that in the mid-1970s the M8 was still in use with several armies around the world.

It was a superb fighting vehicle with an excellent cross-country performance. Its 6×6 configuration was arranged with one axle forward and two rear. The crew of four had ample room inside the vehicle, and the main 37-mm gun was mounted in a circular open turret.

The British knew the M8 as the **Greyhound** but it proved too lightly armoured for British liking, the thin belly armour being vulnerable to mines.

Specification
Light Armoured Car M8
Type: four-man wheeled armoured reconnaissance vehicle
Weight: 7.94 tonnes (7.81 tons)

Every German Panzer division contained a dedicated recce unit. German armoured cars were heavier than their Allied counterparts and were often required to fight for their information.

106

 GERMANY

schwerer Panzerspähwagen SdKfz 234 (8-Rad)

In 1942, German army planners issued a requirement for a new 8×8 armoured car, to be based on the old SdKfz 231 (8-Rad) series but up-armoured. The resulting series was designated **schwerer**

Panzerspähwagen SdKfz 234 and was much lower and more streamlined than the earlier series. The vehicles had thicker armour, greater fuel capacity and a more powerful engine.

The most famous of the range was the **SdKfz 234/2 Puma**, a superb armoured car with a turret enclosing a 50-mm (1.96-in) KwK 39/1 gun. The turret

The Puma was expected to operate well forward of its own troops and fight for information. the M8 was designed for close recce and would gain information by stealth. The difference in their firepower illustrates this: the 234/2 packed a 50-mm anti-tank gun that was effective against most Allied armour, while the 37-mm gun on the M8 was good against nothing bigger than the water truck.

# of World

e M8 had a 37-mm
ain armament and
)-cal co-axial
achine-gun and .50-
l anti-aircraft
achine-gun. This
es not begin to
mpare with the
4/2 and its main
n, but the M8 was
t expected to fight
r information and
ed its gun only in
lf-defence. Also, it
d have some anti-
rcraft protection.

The M8's range of only
563km is just over half
that of the 234/2. The
Germans learned the
value of long-range
recce patrols in the
desert campaigns of
World War II.

Being about a third
lighter than 234/2,
the M8 did not
require such a huge
amount of power
from its engine. The
rival Puma, however,
had far better cross-
country performance.

The four-man crew
(commander,
gunner, driver, co-
driver/radio
operator) is typical
of the manning of
vehicles such as the
M8.

Armament: one 37-mm (1.46-in) gun, one 7.62-mm
(0.3-in) Browning co-axially mounted machine-gun, and
pintle for 12.7-mm (0.5-in) Browning anti-aircraft gun
Performance: maximum speed 89 km/h (55 mph);
range 563 km (350 miles)

Dimensions: length 5 m (16 ft 5 in); width 2.54 m
(8 ft 4 in); height 2.248 m (7 ft 5 in)
Powerplant: one 82-kW (110-bhp) Hercules JXD
petrol engine
Users: USA, UK, many nations post-war

One reason for the lightness of the M8 was its
thin armour: no thicker than 19 mm at
maximum and right down to a mere 4 mm –
hardly even armoured at all. This was a
particular grievance of the British forces, who
were less than impressed by the protection
against mines.

had originally been intended for the Leopard light tank,
which was cancelled, and when reworked for the
Puma the result was powerful enough for the vehicle
to counter the increasing use of light and other tanks in
Soviet army reconnaissance units. The turret had an
excellent ballistic shape and also mounted a co-axial
MG 42 machine-gun.

Other variants of the basic SdKfz 234 were the
234/1 with a 20-mm KwK 30 or 38 cannon: this was
used as a commander's vehicle. It had an open turret,
although it was often screened with wire to prevent

hand grenades from entering. The **234/3** mounted a
short 75-mm tank gun. The final variant was the
234/4, with a Pak 40 anti-tank gun in an open
compartment. This was not successful.

The 234/2's turret
was originally
designed for use on
a light tank; hence
its superior ballistic
shape compared with
that of the M8.

The 234/2 was unique
in having a secondary
driving position. This
was occupied by the
radio operator, who
doubled up as rear
driver. This enabled
fast withdrawal and
equal speeds both
forwards and
backwards.

Specification
schwerer Panzerspähwagen SdKfz 234/2 (8-Rad) Puma
Type: four-crew 8×8 wheeled reconnaissance vehicle
Weight: 11.74 tonnes (11.55 tons)
Armament: 50-mm (1.96-in) KwK 39/1 gun and MG
42 7.92-mm (0.311-in) co-axially mounted machine-gun
Dimensions: length (gun forward) 6.8 m (22 ft 4 in),
length (hull) 6 m (19 ft 8 in); width 2.33 m (7 ft 6 in);
height 2.38 m (7 ft 9 in)
Performance: maximum speed 85 km/h (53 mph);
road radius 1000 km (625 miles)
Powerplant: one Tatra Model 103 air-cooled diesel
engine developing 157 kW (210 bhp)
Users: Nazi Germany until 1945

The impressive V-12 air-
cooled diesel gave an
impressive performance,
despite its weight. The
only problem with diesels
is a tendency to be smoky
and noisy, so they are less
suitable for the recce role.

FROM LANCERS TO LUCHS:

The development of the armoured car

The reconnaissance vehicle has changed dramatically over the years, from a man on a horse with a telescope, to an eight-wheeled amphibious all-terrain vehicle with sophisticated night sights and advanced communications.

Knowing about the enemy has always been a key to winning battles. Commanders with more accurate information than their opponents have often been victorious even when outnumbered. For hundreds of years, that battlefield information was provided by the cavalry, scouting far ahead of the main army.

World War I was the first industrial war, and there was no room in it for horsemen. Armoured vehi-cles were used in their place from the earliest days of the conflict. Belgian armoured cars harried the advancing Germans, and the Royal Navy was quick to follow suit with its Lanchesters and Rolls-Royces. As the war settled down in the trenches, however, there was no place for these converted commer-cial vehicles, and many were moved to other, more open fronts. In Russia and in the desert, the armoured car performed the classic

40

World War II

By the outbreak of World War II, the armoured car had developed into purpose-designed vehicles rather than being modifications to existing vehicles. However, basic attributes remained constant. Armoured cars had to be light and fast to obtain the timely information the commander required.

The chosen form was typically 4×4, as with the British Daimler scout car, although the Germans quickly moved to 6×6 and ultimately 8×8 vehicles, resulting in the excellent SdKfz 234 series. In fact, many of the vehicles in service today show a very pronounced similarity to the vehicles of 50 years ago.

Compare the difference in size between the German SdKfz 231 (above) with the British Humber armoured car (left). The German trend towards large, complicated vehicles was already well set. The British, on the other hand, preferred to go for much smaller and manoeuvrable vehicles. This is a trend that is still very much in evidence today.

SdKfz 231

Below: The schwerer Panzerspähwagen 231 was not a huge success cross-country, being overweight and underpowered, but it had acceptable road performance.

An M8 in Germany, 1944, passes a knocked-out StuG III assault gun. The M8 was unlikely to have had any part in its demise: its 37-mm gun was far too light.

Above: The Belgians were, in many ways, the inventors of the armoured car: the Minerva came about due to lack of soldiers to fight the Germans. The Belgians took to their cars, and before long added guns and armour.

Combat Vehicle Reconn

Scorpion and Scimitar are the armoured reconnaissance vehicles of the British Army. Every infantry and tank regiment, as well as the special armoured reconnaissance regiments, uses these vehicles. In a tank regiment, Scorpion is the recce vehicle; in the infantry, Scimitar. The armoured recce regiments have both, with a total of 32 Scorpions and 40 Scimitars divided into mixed squadrons of both vehicles. The Scorpion is also in service with the RAF Regiment; each squadron has six vehicles operating in pairs, for airfield defence.

The mobility of Scorpion and Scimitar is an important design consideration. A Lockheed C-130 Hercules can carry two in its hull, and a single vehicle can be carried as an underslung load by Chinook or inside a standard commercial container.

Engine and transmission

Under the armoured plating is a Jaguar Xk 4.2-litre petrol engine with Solex carburettor and electronic ignition. Petrol engines produce less smoke and noise than diesel, so are ideal for the recce role, although much more inflammable. The gearbox is the TN 15 Crossdrive, with seven forward and seven reverse gears. It also allows the hull to do a pivot turn that drives each track in a different direction, thus turning the vehicle on its own axis. Very useful for moving in tight areas.

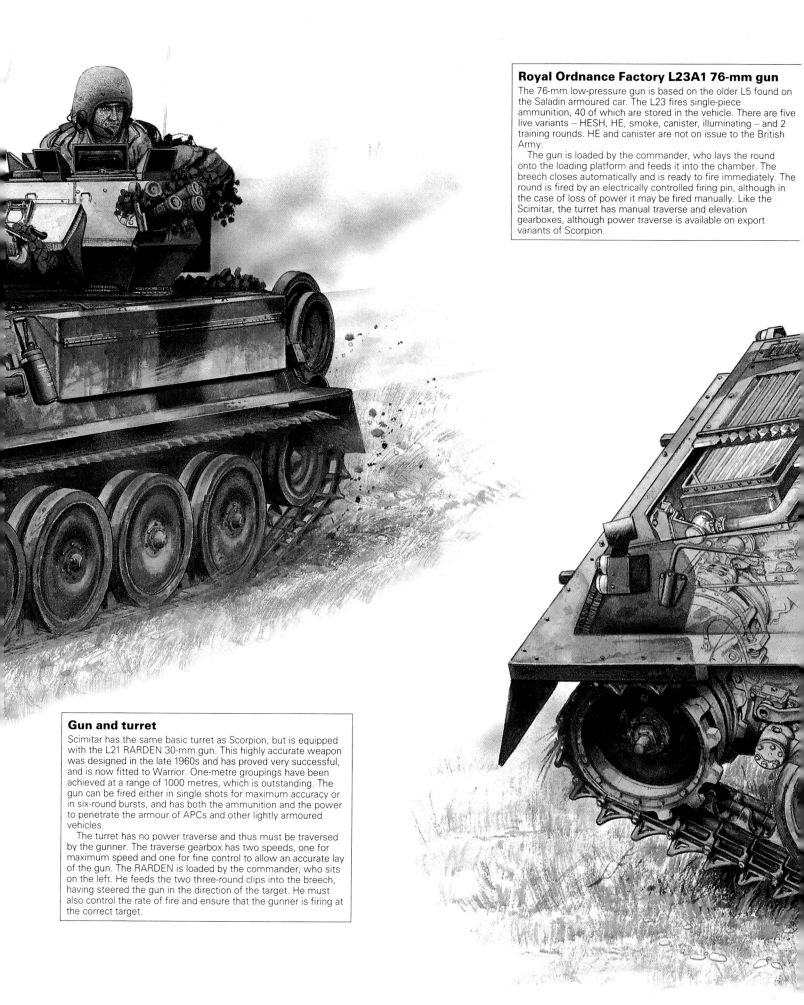

Royal Ordnance Factory L23A1 76-mm gun

The 76-mm low-pressure gun is based on the older L5 found on the Saladin armoured car. The L23 fires single-piece ammunition, 40 of which are stored in the vehicle. There are five live variants – HESH, HE, smoke, canister, illuminating – and 2 training rounds. HE and canister are not on issue to the British Army.

The gun is loaded by the commander, who lays the round onto the loading platform and feeds it into the chamber. The breech closes automatically and is ready to fire immediately. The round is fired by an electrically controlled firing pin, although in the case of loss of power it may be fired manually. Like the Scimitar, the turret has manual traverse and elevation gearboxes, although power traverse is available on export variants of Scorpion.

Gun and turret

Scimitar has the same basic turret as Scorpion, but is equipped with the L21 RARDEN 30-mm gun. This highly accurate weapon was designed in the late 1960s and has proved very successful, and is now fitted to Warrior. One-metre groupings have been achieved at a range of 1000 metres, which is outstanding. The gun can be fired either in single shots for maximum accuracy or in six-round bursts, and has both the ammunition and the power to penetrate the armour of APCs and other lightly armoured vehicles.

The turret has no power traverse and thus must be traversed by the gunner. The traverse gearbox has two speeds, one for maximum speed and one for fine control to allow an accurate lay of the gun. The RARDEN is loaded by the commander, who sits on the left. He feeds the two three-round clips into the breech, having steered the gun in the direction of the target. He must also control the rate of fire and ensure that the gunner is firing at the correct target.

World War I

The first use of armoured cars came in the opening stages of World War I. The Rolls-Royce Silver Ghost armoured car was a straight modification to the staff car that the British Expeditionary Force took with it into Europe. This was to be typical of the first armoured cars, which were often little more than commercially available vehicles with armour and a gun bolted on. Typical of such adaptations were the French Peugeot and the joint UK/Czarist Russia Austin-Putilov.

The trench warfare of World War I precluded the use of lightly armoured mobile vehicles, but the opening of the front in Arabia provided ideal terrain. The exploits of Captain T. E. Lawrence are legendary: his use of just a few Rolls-Royces and other cars caused havoc to the occupying Turkish forces.

With a 37-mm gun and 5.5-mm armour, the French Peugeot armoured cars were well protected and equipped for their time. Later variants had a co-axial machine-gun.

PEUGEOT

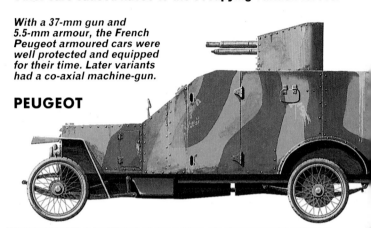

their own initiative in spectacular raids, but their primary function remained that of reconnaissance. Their success was not measured in casualties inflicted or sustained, but in the quality and accuracy of the intelligence they were able to pass back to the rear.

Blurred role

Modern reconnaissance vehicles no longer have such a clear-cut role. Many of the cavalry tasks, particularly the gathering of information, are being carried out more effectively by other means, such as aircraft. Some countries have continued with the development of wheeled and tracked reconnaissance vehicles, while others see little need for them. Armament varies from 20-mm cannon right up to 105-mm tank guns, although there is a school of thought which says that if you give a reconnaissance crew heavy armament they might forget their primary intelligence mission and be tempted to engage the enemy.

However, in the armed forces of many Third World countries the light armoured vehicle is often the prime offensive weapon, and it is for them that many of the heavily armed vehicles have been produced.

Right and below: These are two examples of the early armoured car – they were, literally, cars. Right is a Rolls-Royce armoured car used in the Middle East and made famous by Lawrence of Arabia. Below, German troops examine a captured Austin-Putilov. This was a joint venture between Austin Cars of England, which supplied the chassis, and Putilov, which added the armour and armaments as required.

Above and inset: From the Deccan Horse in 1916 to the Spähpanzer Luchs. Although the change in equipment is, to say the least, dramatic, the purpose has remained the same – the gathering by covert means of timely and accurate battlefield information for use by the commander.

cavalry tasks, scouting out the enemy and raiding behind the lines.

World War II saw the role of the reconnaissance vehicle expand with the advent of mechanised warfare. Armoured cars varied in size from the tiny scout cars like the Daimler Dingo to large, sophisticated and powerful eight-wheelers like the German SdKfz 234 Puma.

They ranged far and wide. Faster and handier than tanks, armoured car units were able to use

Post-World War II

Countries experienced differing degrees of success with their cars during World War II, and this has moulded national style. The Germans were evidently very happy with the Puma, and the later Bundeswehr-dedicated recce vehicle, the Luchs, is clearly a direct descendant. The British pursued the trend of the Daimler and Humber scout cars, through the Saladin and Gerret up to the Fox. However, they, like the Americans, have also opted for a tracked recce vehicle, which although giving excellent cross-country mobility must be questionable in terms of speed and range.

And what of the future? Although satellites and other electronic methods will develop, there will always be the need for battlefield recce. Perhaps the US LAV indicates the current thinking, but remote-reconnaissance devices may soon overtake such vehicles.

ALVIS SALADIN

Introduced in 1956 in an accelerated programme for the Malayan campaign, the Alvis Saladin was, for a long time, the mainstay of British recce. Production of the Saladin finally ceased in 1972. It was a popular vehicle for the crew, being spacious, and for export: it was sold to at least 18 other countries.

The Daimler armoured car was designed in 1939, and was very advanced for its time. It had power steering, disc brakes on all four wheels, and semi-automatic transmission. Seen here in Malaya, it remained with the British Army into the 1960s, seeing action in many conflicts.

Two US light tanks used for recon were the M55 Sheridan (below) and the M41 Walker Bulldog (right). The M41 is used by Denmark as its main recon vehicle and, despite its age, is reliable and well suited. The Sheridan was an experiment that did not go too well. It was widely used in Vietnam but with little success, being far too light, unreliable and poorly protected. It was especially prone to damage from mines.

Communications equipment

The standard radio equipment on the Scorpion is the Clansman VRC 353 VHF radio. This powerful set has the ability to transmit on over 1,840 frequencies, and under ideal conditions can be received over 30 km away. However, used tactically the minimum power setting is used to reduce the chance of detection and interception by the enemy. Two sets are fitted, which allows each vehicle to transmit and receive on both the battle group command net and their own special recce troop net. This allows the troop to operate without taking up air time on the very busy and very important command net. The antenna is the standard one-metre whip antenna, but it is common practice to use only half-metre whip which, although reducing the range, reduces likelihood of the antennas giving the vehicle position away.

Turret

The turret is made of aluminium and is designed to offer small-arms protection. The commander on the left has a x10 roof-mounted No. 52 Mk 1 sight and seven x1 periscopic sights. The gunner has a No. 75 Mk 1 gunner's sight and three periscopes. In addition, there is a Rank Precision Industries image intensifier passive night sight with a magnification of x5.8 and a low magnification of x1.6. The sensor is protected by an armoured cover and is mounted to the right of the gun. To the left is the co-axially mounted L43A1 7.62-mm machine-gun. Three thousand rounds are carried.

External stowage

One of the problems with the original Scorpion was a lack of external stowage. A characteristic of British Army vehicles is how many unofficial 'bins' can be fitted on to carry as much extra kit as possible. Turret and hull back bins are standard, but extra boxes are often fitted to both the turret and the front of the vehicle.

Tracks and suspension

Scorpion is steered by locking one track and allowing the other to continue, thus turning the vehicle. Ground clearance is 0.356 m (1 ft 2 in) and vertical obstacle clearance is 0.5 m (1 ft 7 in). The tracks exert less pressure than a man, and so allow Scorpion to cross areas that foot soldiers cannot.

SAS LAND ROVER 'PINK PANTHER'

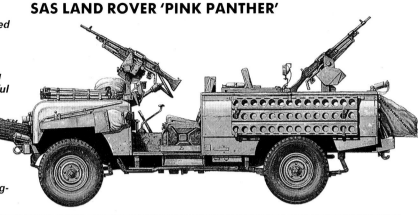

Left: The Soviet BRDM-2 was used to ferry Iranian prisoners during the Iran/Iraq war. The BRDM was a typical Soviet vehicle: cheap to produce, easy to maintain and simple to operate, making it ideal for export. Although not that useful in the middle of the desert, the BRDM is fully amphibious.

Right: Bristling with guns, the SAS 'Pink Panther' long-wheelbase Land Rover vehicle is the recce vehicle for the SAS long-range desert patrol group.

The Bradley M2/M3 Fighting Vehicle is state-of-the-art stuff. Equipped with a 25-mm Chain Gun and the TOW missile, there is no difference between the cavalry M3 and the infantry M2, but inside, the M3 carries extra TOW missiles and fewer crew. It is questionable whether this vehicle is well suited to its recce role, being tracked, heavy and tall.

SPÄHPANZER 2 LUCHS

With its well-deserved reputation for excellent cross-country performance, mobility and, above all quietness, the Luchs is the definitive recce vehicle. It has two drivers, one forward and one rear, and a crew of four. It can do the same speed forward and in reverse. The turret, which has power traverse, has a 20-mm cannon and co-axial 7.62-machine-gun. The Luchs is fully amphibious, has nuclear, biological and chemical protection filters, and excellent night and day sights.

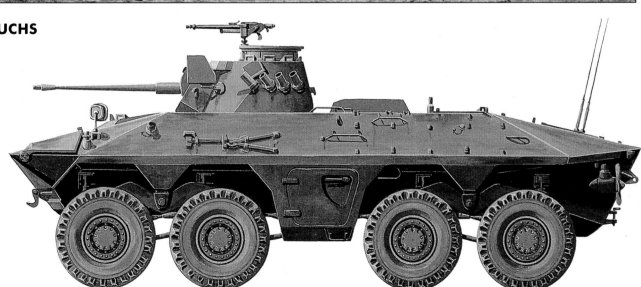

TRACKS in EUROPE

The 5th Royal Inniskilling Dragoon Guards were in the follow-on forces after D-Day. Their recce squadron was typical of the time.

The German forces were hard-pushed from D-Day, 6 June 1944, to VE Day, on 8 May 1945. Once the Allied invasion had taken a firm hold the German retreat was complete. One of the tasks of recce is to ensure that constant pressure is put on a retreating army: if not, they may have time to reorganise and mount a counterattack.

Left: The Stuart light tank, known as an 'atrocity on tracks', was little loved by the British cavalry who had to use it for recce. They later converted it to the 'sawn-off' mode by removing the turret and 37-mm gun, mounting a pintle for the machine-gun in its place.

In World War II recce was recognised as a vital tool, and every armoured regiment had not a recce troop but an entire squadron. The niceties of a peacetime army allow a well-balanced and well-thought-out organisation, but in the realities of war you make do with what you can. The 5th Royal Inniskilling

and objectives, plus a full set of codes and code words.

"On one occasion, not long after landing in France, 1st Troop were sent out to check that the route was clear. As they advanced, the troop leader led the patrol at speed to a small castle by a river crossing. Near the castle was a small farm which the troop took the precaution of clearing first before clearing the castle itself. The tanks and the scout car parked in the courtyard of the farm and set off on foot to clear the farm. As they did so there was an almighty explosion and a hail of fire. Germans in the castle had fired a Panzerfaust at the car, but fortunately missed.

"The tank crew made a dash for it, and despite the fire managed to remount. The 37-mm gun did little to the 3-foot walls of the castle, and the patrol was running short of ammo. The carrier crew, under the cover of the tank, made its escape with the tank. The problem remained that the scout car, though hit by small-arms fire, had not brewed up and a full set of codes was on board.

"1st Troop rearmed and went forward to see if recovery was possible, but came under

withering fire. It was clear that the Germans behind the ramparts were able to bring highly accurate fire down on any rescue attempt. The only thing to do was destroy the car before the Germans could get to it. 1st Troop kept it covered all day until night fell. Under the cover of darkness and unseen by the Germans in the castle, a sniper was brought up. His first round went straight through the petrol tank and the car brewed up. We never made that mistake again.

Push from Normandy

"During the push from the beach-heads of Normandy that would eventually take the Allies to Berlin, the Germans formed themselves into small pockets of harassing infantry. It was just getting light as 1st Troop reached their objective of the night before. They pushed on with orders to report on whether it was held so that the advance could continue. The lead Honey came across a sleepy German who paid for his late rise, but the firing woke the entire defence. Flares went up and many Germans were seen. The road was too narrow and hedged so thickly on each side

Above: The trouble with recce vehicles was that they were thinly armoured. Weapons like the German light 5-cm mortar were a threat. You had to take out mortar positions before they took you!

The Stuart Tank

Dragoon Guards, an armoured regiment that had been at the retreat from Dunkirk, was sent in after the D-Day landings to take up the advance.

"Recce squadron had a nasty shock. It took the form of the light 'Stuart' tank. The semi-official name for this atrocity-on-tracks was the 'Honey'. The gunners who fired the 37-mm cannon thought little of their chances against anything tougher than a water truck. The drivers, used to the powerful Rolls-Royce engines of the Cromwell, weren't too keen. Recce squadron now formed into three troops each consisting of one tank, one Bren gun carrier and one Daimler scout car.

Brewed up

"One important lesson learned on the advance was the necessity to guard information. Recce squadron, since it operated direct from HQ, had maps marked up with all the regiments' positions

The US M3A1 light tank was introduced into the British Army in the later stages of the World War II. The basic armament was a 37-mm gun with a co-axial 7.62-mm machine-gun and three other 7.62-mm guns (one on the turret roof for AA defence, and two fixed in the sponsons for use by the driver). Armour thickness ranged from 15 mm up to 43 mm. When in use by the British it was referred to as the Stuart, but acquired the nickname 'Honey'. Although a reliable tank, it was felt by the British troops to be too poorly armed and underpowered.**

Although shown here in its US colours, the M3 light tank was used by a variety of Allied troops when it was officially called the Stuart.

that it was decided to try to find another way into a place that was less favourable for bazooka men and snipers.

"Accordingly, the troop pulled out and came in from the other side. It was quiet. The Honey led, followed by the carrier. A Frenchman stood at the gate and, as he started to speak, the Germans opened up on the leading tank. It was impossible to turn, so we sped through the village. As we shot through, a Panzerfaust gunner opened up, but the round fell just in front of the Bren gun carrier. The troops were now through the village and behind the Germans.

Dash for it

"The carrier driver spotted two Germans making a dash for it across the field and the carrier took up pursuit, firing madly, but as the Germans reached the edge of the wood the carrier came under heavy and sustained fire. It withdrew under fire. The troop radioed back the information to HQ.

"The advancing tanks had organised with the infantry, and at about three in the afternoon the attack went in. It was a complete success. Some 150 prisoners were taken. With the enemy cleared, the division was able to continue its advance with the sure knowledge that the route was clear.

Above: The versatile Panzerfaust was a very efficient and cheap weapon. It could smash through a recce vehicle's armour with ease.

"The following morning we were on the advance again, leading the regimental column, and were soon crossing the Somme. Recce went ahead and we had identified a bridge intact about a mile from the city of Amiens. By night we were across, and the rest of the regiment was preparing to cross at dawn. The route, to begin with, lay along a straight road lined on each side with poplar trees, but quickly entered a thick wood. The wood was thicker with Germans than trees. 1st Troop kept the

wood under observation while the others scooted around to clear another route.

"The new road to be followed ran alongside the side of a hill with pasture above and close wooded country below. Halfway along, the lead carrier, suspicious of the movement in a farmhouse in the wood, stopped. As the gunner of the Honey brought his gun to

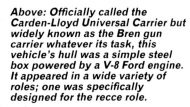

Above: Officially called the Carden-Lloyd Universal Carrier but widely known as the Bren gun carrier whatever its task, this vehicle's hull was a simple steel box powered by a V-8 Ford engine. It appeared in a wide variety of roles; one was specifically designed for the recce role.

Left: A fine piece of German engineering, the 5-cm Pak 38 was a well-crafted gun. It was simple to use, had a very effective tungsten penetrator round and, by use of light alloys in its construction, was easy to move.

Right: The finest tank of its time, the Tiger mounted the 88-mm gun, which outclassed anything the Allies had. Its frontal armour was impervious to their guns at anything but the closest ranges. Its only fault was its excessive weight, which made it unmanoeuvrable. If it could be attacked from the side or rear, the Tiger brewed up like any other.

EYE WITNESS

"The advance from the Seine had been so swift that there was a shortage of maps for the country ahead. On the morning of the 5th the regiment headed for the Escault canal and the bridges at Melle and Wettern. The Melle bridge was found to be blown, but our patrol sent to the Wettern bridge found it intact. It was a wooden affair with sections that could be lifted in drawbridge fashion to allow vessels on the canal to pass. This section was under fire at the time. But here the Belgian underground movement came to the rescue. Operating jointly with our recce patrol, they managed to get the drawbridge lowered. The entire regiment as well as the 5th Royal Tank Regiment were able to get over. This led directly to the liberation of Ghent.

Sergeant, Armoured recce squadron, 1944

bear, one of the three well camouflaged 50-mm German anti-tank guns opened up and hit the turret.

Complete success

"The German gunners turned their attention to the leading vulnerable carrier and brewed it up. As the crew evacuated they were machine-gunned by a Spandau with the 50s. The 3rd Troop went out again to check out a higher route out of range of the 50s, but once more they were spotted by the enemy and engaged as they set off. The recce squadron leader, in his half track, came up, very much aware that the entire regimental advance was being held up. A sniper put a round through his windscreen and out through the canvas hood. He withdrew quickly.

"There was nothing for it but to call up the tanks and attack the position, although without supporting infantry this was a very dangerous operation and one that may well be very heavy in casualties. Meanwhile, 3rd Troop had put in a call for the 7.2-inch guns, which opened up with devastating accuracy. 1st Troop had withdrawn and picked its way through, avoiding contact, and by careful route selection and driving got themselves well forward.

"Just as they broke the crest of the hill they came across the rear of the German position, where the Germans were hurriedly digging in and setting up a mortar position. The stage was set for a grand shoot in the valley under the conditions usually met only on range practice. The regiment advance continued quickly without the need to attack. It was a complete success."

CONTACT, TANK, WAIT OUT

The eyes and ears of the battle group are the recce troop. It is the first into battle and the last out. The success or failure of a battle group depends totally on the quality of the information it has from recce.

The skill of the recce troop or platoon is perhaps one of the hardest to master. It requires stealth, extremely accurate reporting skills, very detailed observation, and courage. Recce units are the first into battle and the last out.

Recce tactics are task-orientated: they vary with the job in hand. But they can be divided into two main areas – recce in the advance/assault, and recce in defence/withdrawal. It is worth remembering that a recce troop is one of the most versatile assets in a battle group and may be called on to carry out an unbelievable range of additional tasks, from demolitions to route-marking.

There is a difference in tactics between British recce and other countries. British forces have always conducted 'recce by stealth' – they try to get the information without the enemy knowing they are there. Other countries, such as the Soviet Union and West Germany, conduct 'recce by fire' – firing at the enemy and seeing what fires back! Here we look at British tactics.

Right: A column of T-64s advances along a forest track. A sighting such as this is exactly the sort of vital information that commanders need: information on strengths, intentions and locations of the enemy must be passed back as soon as possible to be of any use. Your 76-mm gun is for defence only – try engaging this lot and see what happens.

Left: Bounding overwatch is a basic principle of warfare – vehicles do not move unless they are covered by the guns of other vehicles in 'overwatch' positions.

Engineer recce (in the Spartan) always travels with battle group recce in the advance. It will be able to report on enemy obstacles and minefields. Being this far forward means that it is well positioned to conduct detailed recce and call forward the vehicles and men needed to tackle the problem.

1 Bounding overwatch

Recce advances about 1000 m in front of the tanks, which cover its movement in case of contact, when the tanks will engage and destroy the enemy. The eight-car recce troop is split into four sections of two cars. Tanks advancing on a wide front will have the troop deployed with all four sections operating right across the front. If the tanks are advancing on a narrower frontage, it may be wise to keep one recce section in reserve in case it is needed for a special task. When the enemy is contacted recce troop will conduct a sweep to investigate if it is the only position across the front. If so, then all the sections will investigate. If there is more than one position, a section will be used to keep one position under observation while the other is attacked. Unless the positions are very small, they will not be able to be taken on simultaneously.

A well-cammed-up Scorpion picks its way through an urban environment. The problem for any armoured vehicle in such terrain is that it is very vulnerable to attack. It would be better to scoot round the edge of the town, or if this is impossible you must take a cautious route.

Aim is to see without being seen

Conduct probing recce of enemy position

One vehicle keeps enemy under observation

Above: Once contact is made, one car keeps the position under observation while the others conduct probing recce of the position to establish enemy size and strength.

2 Sighting

Once contacted, the enemy must be kept under observation at all times. At least one car will do this, and it may be better to use a section. The rest of the troop will conduct a probing recce of the enemy to establish the exact size and extent of the position. Recce troop will also be used to check out routes to the position and also to look for a suitable area from which to launch the attack. Other tasks may include bringing down controlled artillery fire and smokescreens, and they may have to carry out engineer recce if the engineers are engaged elsewhere. Such tasks could include minefield marking, investigating anti-tank obstacles and checking out bridges etc.

Where possible, avoid taking obvious tracks. If the enemy has had any time he will have laid off route mines to destroy any armoured vehicles coming up them. Scorpion is highly mobile – go cross-country.

TACTICS

Above right: The art of camouflage is to break up the outline of the vehicle. When in position and closed down, this well-cammed Scorpion would blend perfectly into its background.

3 Observation posts

The attack on the enemy has gone well, and the battle group commander has consolidated and gone into defence. The main function of recce is to provide the commander with timely battlefield intelligence. There is no phase of war in which this is more important. Recce will go forward to set up a screen: a line of observation posts tactically situated to provide the maximum possible observed area and cover all likely lines of approach. Along this screen, recce troop will set up four two-vehicle OPs. Although it would be better to have eight single-car OPs, with only a three-man crew in each car, fatigue would be a real problem. OPs will have a system of resupply thought out in advance — there is no point in having a brilliantly cammed-up OP and then letting the resupply vehicle drive up next to it to deliver food. The normal technique is to arrange a 'dead letter box' where the resupply vehicle will drop off the goods, and one man from the crew will sneak out to pick it up. It is a laborious task, but essential if the OP is not to be compromised.

Scorpion in forward screen
Task: to provide early warning of enemy movements

Right: In defence, recce will put out a screen of OPs to observe and report on enemy movement. It will be well cammed and will always operate in pairs to ease the problem of fatigue.

4 Withdrawal

The extraction of a fighting force in contact is one of the most difficult phases of war to control. Withdrawal requires speed and aggression and the ability, should the opportunity present itself, to mount very fast counter-attack. Who will provide the commander with the information on when to do this? Recce troop. Recce will be the last to withdraw; the first will be the infantry in their APCs, and their move will be covered by the tanks. With the infantry out, the tanks will withdraw, at speed, hulls in the direction of travel, gun pointing at the enemy. As always, it is important that one foot is kept on the ground at all time; the tanks must not all move at once. Recce will keep the enemy under observation and continue to report up to the last minute. When it withdraws, it goes at maximum speed, ensuring that when one vehicle moves the other in the section is covering his move.

Left: Although no longer capable of swimming, the Scorpion can deep-wade – perhaps not always that popular a prospect with the driver! The Scorpion exerts less ground pressure than a human foot, and this enables it to go places where even dismounted recce patrols cannot. By careful route selection a recce patrol can outflank an enemy position that may have considered some routes impassable to armour and infantry – forgetting the fantastic mobility of the CVR.

Below: In the withdrawal the infantry moves first, covered by the armour. Recce troop will remain in position to the last possible minute, observing the enemy, ensuring that contact is not lost. During this operation it is vital that the commander knows exactly what the enemy is doing, and only recce can give him this information. Again the recce patrols are made of two vehicle teams, one always covering as the other moves.

One vehicle covers, others move

APCs move out first

Guns face enemy; hulls face direction of travel

Recce maintains observation while tanks withdraw

Left: A high-speed withdrawal. The gun is pointed at the enemy and the hull towards the direction of travel. Although this exposes the weakest armour, it allows a faster retreat. The gun is pointed in the enemy direction for two reasons: to shoot them, and to stop you from being shot by your own side. If they saw a tank moving at speed with its gun pointing at them, they would shoot first and ask questions later.

MAIN BATTLE TANKS

Sixty-tonne metal beasts smash through anything in their path, cutting a swathe of destruction. Or are they just metal pillboxes, vulnerable to the single infantryman and a cheap throwaway weapon?

In the uproar of artillery, grenades and gunfire, the sound of the tanks' Daimler engines was hardly noticeable. The machines ran along slowly, oblivious, sponsons on each side spitting machine-gun bullets or six-pound rounds. To the men facing them they were nothing short of monstrous; creeping from the smoke, like houses come to life, crashing through wire and over trenches, crushing everything before them at a steady two miles per hour.

All around was bare earth. Everywhere were the stinking remnants of a stagnant war: weapons, equipment, bodies. For the men of C and D Companies, Heavy Section, Machine-Gun Corps, there was nowhere to go but forward, through the smoke and the dust and the debris, breaking the defensive trench lines and sending terrified men running for their lives.

On the Somme, in the 10 weeks of battle that took place before the tanks arrived, the Allies had measured their gains in yards and their losses in hundreds of thousands of men. A way had to be found to make more economical headway against an enemy dug in and concreted over, armed with machine-guns effective out to 1,000 yards.

Above: Arguably the best tank in the world, the Leopard 2 is highly mobile with excellent cross-country mobility and the Rheinmetall 120-mm smoothbore gun – one of the world's finest. It has only one fault, and that is its questionable armour protection.

Right: Unlike most tanks, which have a conventional diesel engine, the US Army M1 Abrams has a gas turbine that gives a lot more power from a lot less weight. However, it consumes huge amounts of fuel.

The solution was the armoured vehicle, a sort of pillbox that could cross trenches and broken ground under fire, preparing the way for the infantry while keeping its occupants safe. But only if it could function and keep moving across a terrain devastated by the ravages of both war and winter. Guderian, Hitler's genius tank commander, credits the tank with winning World War I for the Allies. The truth is less clear-cut, but one thing was certain, even then – the god of war had put on a new mask.

Shock value

Through the 1920s, the trend was towards lighter tanks with better armaments, higher speeds, wider range and smaller crews, and it was these characteristics that coloured the thinking of the new breed of tank strategists that was emerging. Why should tanks be used simply to support the infantry, and thus be limited to perhaps two miles per hour? Why not exploit the shock value of this huge metal beast, appearing without warning in all the most vital places – places that had never been vulnerable before – destroying men, munitions and materiel and then moving on?

And that was not all. The notion of the armoured column, crushing all before it, only needed magnifying to become a truly new way of waging war – Blitzkrieg!

Two men's names are permanently linked

The Lockheed Galaxy C-5 transport aircraft can carry two M60s in its hold. For political and strategic reasons, this is a useful ability to have at your disposal

with the development of the new strategy. Basil Liddell Hart was a junior officer in the British Army. In 1920, he re-wrote the infantry manuals in the light of the experiences of the spring offensive of two years before. By 1927 his ideas had found a measure of acceptance. Despite being restricted, his pamphlet *Mechanised and Armoured Forma-*

tions came into the hands of the German High Command, who had at least one officer who recognised its worth – Heinz Guderian, in the opinion of many the finest tank officer the world has yet seen.

The Anschluss that incorporated Austria into the Third Reich; the annexation of Czechoslovakia; the invasion of Poland; and finally the Blitzkrieg proper, that saw the collapse of France in little more than a month, all proved Guderian's theories. It seemed that tanks and motorised infantry, used aggressively and supported by ground-attack aircraft, were unstoppable. And then, on Midsummer's Day 1941, Hitler tore up the non-aggression pact he'd made with Stalin, and sent the Blitzkrieg against Russia.

The Eastern Front was to produce World War II's major tank battles. For two years,

How the tank got its name

The term 'tank' had yet to be coined in 1915. The experimental vehicles were known as Landships (and first went into action as such, graced with such titles as HMLS *Kia Ora*), but to confuse the enemy a handy name that didn't betray their function had to be found. They tried 'cistern' and 'boiler' before settling on 'tank', which raises interesting possibilities. The Royal Boiler Regiment? Anti-cistern guided weapons?

Ever since the first important use of the 'tank' at Cambrai on the Western Front in World War I, the General Staff have been unable to cope with the thinking that its speed forces on them.

MAIN BATTLE TANKS Reference File

129
T-80

FORMER USSR

The **T-80** is one of several MBTs in service with the tank formations of the Russian army and, in essence, is a major development of the T-64B with several improvements as well as features to eliminate some of the T-64's tactical shortcomings. The most obvious changes are a glacis of laminate-type armour, a dozer blade under the hull nose, a revised turret with an inner layer of 'special armour', new suspension with rubber-tyred rather than resilient steel road wheels and, perhaps most importantly of all, a gas turbine powerpack in place of the T-64's troublesome 560-kW (750-hp) five-cylinder diesel engine.

The T-80 has the same main armament as the T-64, though probably in a later mark with an improved fire-control system. This gun can fire the AT-8 'Songster' anti-tank/anti-helicopter missile, as well as three types of separate-loading conventional ammunition. Other items includes standard features such as an NBC system and night vision equipment. The type has been considerably updated in service, the most important such aspect being the retrofit of reactive armour, which is intended to reduce the effect of incoming anti-tank missiles' HEAT warheads.

Specification
T-80
Type: three-crew main battle tank
Weight: 42 tonnes (41.36 tons)

Armament: one 125-mm (4.92-in) smooth-bore gun with 40 rounds, one 7.62-mm (0.3-in) co-axial machine-gun with 2,000 rounds, one 12.7-mm (0.5-in) AA machine-gun with 500 rounds, and normally nine smoke-dischargers (five on the left and four on the right sides of the turret)

Performance: maximum speed 75 km/h (46.6 mph); range 600 km (373 miles) with external long-range tanks
Dimensions: length overall 9.9 m (32 ft 5.75 in); width 3.4 m (11 ft 1.9 in)
Powerplant: one 735-kW (985-hp) gas turbine
User: USSR

Above: The T-72 is a popular Russian export model, not only in the former Warsaw Pact countries but throughout the world. The 125-mm gun is smoothbore and, unlike the weapons on most Western tanks, has an autoloader. Service T-72s have additional armour and a superior engine and sighting equipment.

Left: Tanks, unlike dug-in infantry, do not wait for the battle to come to them, but go and look for it anywhere on the battlefield. They need commanders who can think fast and act quickly and decisively if the tank is to be used to its full aggressive potential.

130

ITALY

OTO Melara/IVECO C-1 Ariete

In 1984 IVECO (Fiat) and OTO Melara agreed on the joint development of an MBT with a 120-mm (4.72-in) gun and an 8 x 8 wheeled tank destroyer with a 105-mm (4.13-in) gun. IVECO has great experience with wheeled heavy vehicles and was therefore the logical choice as prime contractor for the B-1 tank destroyer, while OTO Melara's experience with heavy weapons and tracked chassis is reflected in this company's responsibility for the **C-1 Ariete** (battering ram) tank. The first of three C-1 prototypes ran in 1987, and the type should enter service in the early 1990s.

There is nothing unconventional about the C-1, though its angular external appearance confirms that the basic structure is of welded steel with a layer of composite armour over the frontal arc. The main armament is a smooth-bore gun developed by OTO Melara, and this is stabilised in two dimensions and used with an Officine Galileo TURMS OG14L3 fire-control system. This latter includes stabilised optical and thermal sights, a laser rangefinder and sensors for ambient conditions all feeding data into the digital central computer that produces a high-quality solution to the fire-control problem.

Specification
OTO Melara/IVECO C-1 Ariete
Type: four-crew main battle tank
Weight: 48 tonnes (47.24 tons)

Armament: one 120-mm (4.72-in) smooth-bore gun with about 40 rounds, two 7.62-mm (0.3-in) machine-guns (one co-axial and one AA) with about 2,500 rounds, and four smoke-dischargers on the sides of the turret
Performance: maximum speed 65+ km/h (40.4+ mph); range 550+ km

(342+ miles)
Dimensions: length overall 9.669 m (31 ft 8.6 in); width 3.545 m (11 ft 7.6 in) over the skirts
Powerplant: one 895-kW (1,200-hp) Fiat V-12 MTCA diesel engine
Users: Italy and Spain

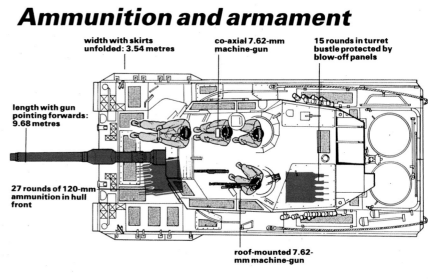

Ammunition and armament

width with skirts unfolded: 3.54 metres

co-axial 7.62-mm machine-gun

15 rounds in turret bustle protected by blow-off panels

length with gun pointing forwards: 9.68 metres

27 rounds of 120-mm ammunition in hull front

roof-mounted 7.62-mm machine-gun

Any fool can be uncomfortable. If you have to live off your vehicle for weeks at a time, anything to make life just a bit more bearable is worth it, even if it is a touch untactical. Deck chairs, perhaps?

the Germans persevered. In the final analysis, the campaign was to be their worst mistake. In July 1943 the tide really turned against them, in the battle for the Kursk Salient.

The armies pitted against each other were huge – 37 German divisions, including 3,000 tanks, against an even larger Russian force.

By the eighth day of the offensive, Soviet tanks and anti-tank guns had destroyed 2,900 of von Manstein's tanks, 195 self-propelled artillery pieces, 1,392 aircraft and 70,000 of his men. The Wehrmacht went no further in the East.

Workable combination

Guderian had isolated three factors that had to combine if the tank were to work as a weapon: armour, firepower, and reliable mobility. By 1943 that combination was work-

In the Leopard 2, most ammo is stowed in the hull for protection with just a few rounds in the turret 'ready round' bins for immediate use. Once these are empty the commander will withdraw from the battle and restow these bins from the hull ammo. The turret ammo bins have blow-off panels in case they are hit.

ing for everyone; the growing pains were over. As tanks on both sides got bigger, faster, tougher and more reliable, the tempo of war hotted up.

By the time of the Normandy landings, in

131

UNITED KINGDOM

Vickers Challenger 2

The bid for a replacement for the Chieftain was between the M1 Abrams, the Leopard 2, and a new and then unproven tank, the Vickers Defence Systems **Challenger 2**. The Challenger 2 was finally selected in the early 1990s.

In many ways the Challenger 1 looks similar to the Challenger 2 – they share a hull and many automotive parts – yet the two tanks are fundamentally different with over 150 improvements in the latter. Most significant amongst these is the turret, which is effectively a new design with updated armour protection. It has the XL30 120-mm gun firing two-piece ammunition and, though the engine is the same as the Challenger 1, the Challenger 2 features

a new TN-54 six-speed gearbox.

The Challenger 2 is a sophisticated weapons platform. Both commander and gunner have stabilised sights; the commander's is French designed and the gunner's laser sight was designed by the British firm Barr and Stroud. It is linked with an improved Thermal Observation and Gunnery System (TOGS) which gives an unparalleled night-fighting ability. The turret also has enhanced NBC protection.

Specification
Vickers Challenger 2
Type: four-crew experimental main battle tank
Weight: at least 62 tonnes (61 tons)
Armament: one 120-mm high-

pressure rifled barrel with at least 60 rounds; one co-axially-mounted 7.62-mm L94A1 Chain Gun with 7,000 rounds; loader's machine-gun with high elevation mount; probably L37 7.62-mm machine-gun with 200 rounds
Performance: not known, but superior to Challenger 1

Dimensions: (approx) length 11.56 m (38 ft); width 3.518 m (11 ft 6 in); height 3.2 m (10 ft 6 in)
Powerplant: one 894-kW (1,200-bhp) Perkins diesel engine plus auxiliary 27.6-kW (37-bhp) Perkins diesel engine
Users: British Army

132

GERMANY

Krauss-Maffei Leopard 2

To succeed its Leopard 1, which had entered service in 1965, the West Germany army planned to adopt the US-German MBT-70. This was cancelled in 1970, but West Germany then pressed ahead with the extremely capable **Leopard 2** using many MBT-70 features and indeed components. The Leopard 2 was accepted in the mid-1970s and entered service in 1979.

Since then the Leopard 2 has proved itself an excellent MBT with a potent combination of firepower, protection and mobility, the last derived from its powerful yet reliable engine and advanced suspension. The fully stabilised Rheinmetal smooth-bore gun fires fin-stabilised projectiles in the anti-

tank role, and is used with a capable fire-control system; this ammunition uses a semi-combustible cartridge case whose only remnant is a brass stub that drops into a collectors bag under the breech and helps to keep the fighting compartment uncluttered. A high level of protection is provided in part by the tank's mobility but mostly by its construction of spaced multi-layer armour, together with steel-reinforced rubber skirts (protecting the upper run of the tracks) and a fast-acting fire extinguishing system.

Specification
Krauss-Maffel Leopard 2
Type: four-crew main battle tank
Weight: 55.15 tonnes (54.28 tons)

Armament: one 120-mm (4.72-mm) smooth-bore gun with 42 rounds, two 7.62-mm (0.3-in) machine-guns (one co-axial and one AA) with 4,750 rounds, and eight smoke-dischargers on each side of the turret
Performance: maximum speed 72 km/h (45 mph); range 500 km (342

miles)
Dimensions: length overall 9.668 m (31 ft 8.6 in); width 3.7 m (12 ft 1.7 in) over the skirts
Powerplant: one 1118-kW (1,500 hp) MTU MB 873 Ka 501 multi-fuel engine
Users: Netherlands, Switzerland and Germany

June 1944, the Blitzkrieg formula was the only game in town, though it was a much more difficult game to play in the small, enclosed fields of Normandy than it had been on the Russian Steppes. A "deadly game of hide and seek", one observer called it, with tanks creeping around the lanes, coming upon each other at point-blank range.

The tanks on which the strategy depended

were getting bigger and better-armed. In 1942, the German Panzer IV tank could pierce 5.6 inches of armour at a range of 500 yards. Two years later, the Tiger II could penetrate twice that thickness at twice the range.

But though the modern tank was powerful, it wasn't all-powerful. Deployed in the wrong sort of country, it was clumsy and vulnerable to anti-tank weapons. Through the jungles of

the Pacific islands and South East Asia, both the Americans and the Japanese consistently used tanks in their World War I role as infantry support weapons – moveable strong-points – and the same was true, to some degree, when the war came to the towns and cities of Germany itself.

But these actions – and even the Battle of

Built as a private venture between FIAT and OTO-Melara, the Italian OF-40 MBT has been produced for export. It is based largely on the German Leopard 1 but with various modifications and fires the standard NATO 105-mm ammunition. A well-trained crew can achieve a rate of nine rounds a minute.

133
Chieftain

UNITED KINGDOM

The **Chieftain** was the main battle tank of the British Army for over three decades. Introduced in 1963 to replace Centurion and the abortive Conqueror, Chieftain underwent 13 updates and it is still in service abroad.

The armament consists of the Royal Ordnance Factory L11 120-mm rifled gun firing two-piece ammunition (round and charge); there is also a co-axially mounted L8 7.62-mm machine-gun with 7,000 rounds and the commander has an L37 7.62-mm machine-gun. Fitted to each side of the turret are six-round smoke grenade dischargers that will lay down a local smoke cover for the tank.

The armour is simple solid steel,

although this has been reinforced with a bolt-on composite turret armour known as Stillbrew. The final modification is the retrofit of the Barr and Stroud TOGS (Thermal Observation and Gunnery System), an outstanding night observation system.

The powerplant is a Leyland L60 13A multi-fuel engine. It was designed to be able to run on any type of fuel, although this was something of a technological dead end, and it now runs exclusively on diesel.

Specification
FV 4030 Chieftain
Type: four-crew main battle tank
Weight: 58 tonnes (57 tons)
Armament: one 120-mm (4.72-in)

rifled bore gun with 52 rounds, two 7.62-mm (0.31-in) guns with 7,000 rounds and six smoke grenade dischargers on each side of the turret
Performance: maximum speed 45 km/h (28 mph); range 225 km (140 miles)
Dimensions: length 10.80 m (35 ft

5 in); width 3.66 m (12 ft); height 2.87 m (9 ft 5 in)
Powerplant: one 634-kW (850-bhp) L60 diesel engine and one H30 Coventry Climax Generating Unit engine developing 17 kW (23 bhp)
Users: Iran, Iraq, Jordan, Kuwait, Oman and UK

134
ENGESA EE-T1 Osorio

BRAZIL

Although not a major arms manufacturer, the Brazilians have produced some very interesting and well-thought-out products. Developed as a private venture, the **Osorio** has been designed for maximum flexibility and with an eye on the export market.

It is one of the few tanks that is designed to have a variety of guns. It can be fitted either with the French-designed GIAT 120-mm smoothbore gun, or with the British 105-mm rifled bore L7. There is also a choice in the machine-guns available: either two 7.62-mm (0.31-in) or one 7.62-mm and one 12.7-mm (0.5-in). Other modifications such as NBC protection are available.

The design philosophy has been to

produce a cheap but effective tank that is simple to maintain and to operate. There is extensive use of modular components throughout, speeding up their repair in the battlefield. The entire powerpack (i.e. engine and gearbox) is removable as one unit in under half-an-hour.

Specification
ENGESA EE-T1 Osorio
Type: four-crew main battle tank
Weight: (120-mm model) 43.7 tonnes (43 tons); (105-mm model) 40.4 tonnes (39.8 tons)
Armament: one 120-mm (4.72-in) GIAT smoothbore gun with 38 rounds or ROF L7 105-mm rifled bore gun with 45 rounds; machine guns of either

7.62-mm (0.31 in) with 5,000 rounds or 7.62-mm with 3,000 7.62-mm rounds and 900 12.7-mm (0.5in) rounds of ammunition
Performance: maximum speed 70 km/h (44 mph); range 550 km (342 miles)

Dimensions: length (120-mm model) 10.1 m (33 ft 2 in), (105-mm model) 9.34 m (30 ft 8 in); width 3.26 m (10 ft 8 in); height 2.89 m (9 ft 5 in)
Powerplant: one 768-kW (1,040-bhp) MWM TBD 234 diesel engine
Users: Brazil

The T-54/55 is in service with more armies than any other tank in the world. One of the reasons for this is its simplicity and reliabilty, not to mention price. Here Libyan T-55s were captured intact when the Chadean forces overran them.

the Bulge – were anomalies. The tank's true terrain was the vast plains of Eastern Europe and the deserts of North Africa, where the tank's exclusive combination of firepower and manoeuvrability gave it mastery over any other fighting strategy.

In the years since World War II, that stark fact has been reinforced. In the two major conflicts – Korea and Vietnam – the tank hardly counted. But in the Arab-Israeli wars and the Indo-Pakistani wars, the tank has shown, once again, that in open country it is

The reality of tank life: track bashing. Most of the everyday life of a tank crewman is spent maintaining his vehicle. As they are so complicated, tanks require constant attention if they are to work when needed.

king – if its armour can stand up to modern anti-tank weapons.

For the modern tank has come to rely more and more on science and technology for its protection. Weapons technology, and new ways of aiming the gun and predicting where the shot will fall, have reached a point of stagnation, and it appears that the next step will be towards guided missiles – or perhaps even missiles that seek out their own targets and home on them, without the necessity of human intervention.

Future developments

The gun still has life left in it yet, though. The modern thinking is towards liquid propellant, metered into the chamber according to the real needs of the next target. Hence a long-range shot will get a full charge of propellant, while a short-range shot will get just as much as it needs. The result is less wastage, and better munitions consumption, so increasing the tank's mission endurance.

And so tank designers have concentrated on the strength of their armour. Chobham Armour – a composite material whose individual components react in different ways to different threats – is the most important step forward, but some put their faith in reactive armour – small plates of explosive, covering the vulnerable surfaces of the tank, and designed to blow up when struck by an incoming projectile, destroying both themselves and it but leaving the tank's armour unhurt.

To the general public, though, the refinements of armour and armament towards which the tank designers are striving mean very little. What they see are the scenes in the streets of Prague, Budapest, Gdansk and Beijing, in which unarmed men, women and children are pitted against 60-ton monsters that can run over them by the dozen and not even register the disturbance. The tank was a terror weapon when it first appeared; that much, at least, hasn't changed.

Combat Comparison

In just 80 years from the first successful tank, the Mk IV, to the latest, most modern and sophisticated battle tank, the US Army M1 Abrams, the tank has been transformed beyond recognition. Its size seems to have remained roughly constant, however, if not its lethality.

135

Tank Mk IV

The **Tank Mk IV** was the most numerous type used in World War I, and benefitted from the lessons learned on the previous three marks. The main differences were improved armour protection (up to 12 mm/0.47 in) and a change in the gun lengths. Original guns had been 40 calibres long (i.e. their length was 40 times their calibre), but the problem with this was that they had a tendency to get stuck in the ground and had to be cleaned before being fired again. The Mk IV's guns were reduced to only 23 calibres.

Living conditions in the Mk IV were, to say the least, poor. The engine was centrally mounted and open inside the hull. It had no suspension so the ride was rough. The crew of eight had no means of communicating with each other apart from banging on the hull.

136

General Dynamics M1 Abrams

In July 1970 development of the MBT-70, a joint US and West German tank, was halted and the US Army then launched a programme to produce an austere version of this vehicle as successor to its M60 MBT. It was soon appreciated that even this downgraded model would be too costly, and in 1972 a completely fresh design competition was launched. This was won by the Defence Division of the Chrysler Corporation (now part of General Dynamics), and the first prototype was completed in 1978.

The 105-mm gun is accurate to over 3000 metres. It is fully stabilised, as are the sights, so allowing it to engage both moving and static targets while the tank is on the move. Even the inert training round would go straight through the Mk IV and out the other side. The co-axial machine-gun has tracer burn-out at around 1000 metres, but can still engage targets at over 1250 metres.

The Mk IV's main armaments were fitted in 'sponsons' on each side of the hull: this was because the track was routed over the top of the hull and so precluded the use of a turret. The sponsons came in two varieties: the so-called 'male', each of which had a 6-pdr gun fitted, and the 'female', which had two Lewis guns in each sponson. Both variants had in addition two more Lewis guns. There was a later variant called the 'hermaphrodite' which, not surprisingly, had one male and one female sponson.

The two 6-pdr guns and two Lewis guns were, for their time, awesome. The effective range was no more than 400 yards and the gun completely unstabilised, making aiming and firing something of a game of chance.

Specification
Tank Mk IV
Type: eight-crew main battle tank
Weight: 28.4 tonnes (28 tons)
Armament: (male) two 6-pdr (57-mm) guns and two 0.303-in (7.69-mm) Lewis machine-guns; (female) six Lewis machine-guns; (hermaphrodite) one 6-pdr gun and four Lewis machine-guns
Performance: maximum speed 6km/h (3.7 mph); range 56km (35 miles)

Dimensions: length 8.05 m (26 ft 4.9 in); width over sponsons 3.91 m (12 ft 7 in); height 2.49 m (8 ft 2 in)
Powerplant: one 78.3-kW (105-bhp) Daimler petrol engine
Users: UK, Italy and Germany (captured ones known as Beutepanzerwagen)

By current standards, the Mk IV would not be considered bulletproof. It was designed to afford the crew bullet and splinter protection but the earlier models had a problem with the bullets splashing through the joins in the armour and burning the crew.

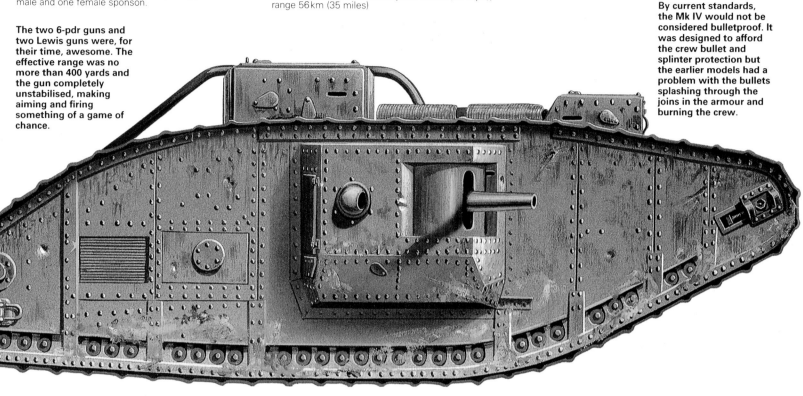

The new tank was accepted for service as the **M1 Abrams**, and production started in 1980. This first model has the same fully stabilised main armament as the M60 it replaced, but in this instance used with a much superior fire-control system with a laser rangefinder, stabilised sights including an imaging thermal component for night-engagement capability, and a digital ballistic computer. Other advanced features are a hull and turret of Chobham-type composite armour, and a gas turbine powerpack for good performance despite its high fuel consumption

and 'thermally visible' exhaust. The **M1A1** version of 1986 introduced the powerful 120-mm (4.72-mm) Rheinmetall smooth-bore gun firing a potent fin-stabilised projectile in the anti-tank role, together with improved vision, protection, fire-control and NBC features, while the forthcoming **M1A2** will have many detail improvements.

Specification
General Dynamics M1 Abrams
Type: four-crew main battle tank
Weight: 54.545 tonnes (53.68 tons)

Armament: one 105-mm (4.13-in) rifled gun with 55 rounds, two 7.62-mm (0.3-in) machine-guns (one co-axial and one AA) with 11,400 rounds, one 12.7-mm(0.5-in) AA machine-gun with 1,000 rounds, and six smoke-dischargers on each side of the turret
Performance: maximum speed 72.5km/h (45 mph); range 498km (310 miles)
Dimensions: length overall 9.77 m (32 ft 0.5 in); width 3.65 m (11 ft 11.75 in)
Powerplant: one 1118-kW (1,500-shp) Textron Lycoming AGT-1500 gas turbine
User: USA, Egypt, Saudi Arabia, Kuwait

The frontal turret armour of the M1 is made up of the British-designed Chobham armour, which is impenetrable to all but the most modern rounds. The M1A2 is further up-armoured with a depleted uranium-based armour (non-radioactive).

The crew of the Abrams can operate in a filtered air environment that will protect them from chemical weapons. There is an integral radio and intercom harness that allows all crew members to not only speak to each other but also to operate the radios.

What is so special about a tank? How does is differ from other armoured vehicles? Simple – when used properly, it is unstoppable.

SHOCK ACTION!

What makes a tank a tank? Well, for a start it has a gun – **firepower**. You can move that gun around the place under its own power – **mobility**. Finally, the crew has some sort of armoured casing around it – **protection**. All major battle tanks have these factors.

Different countries' experiences have caused them to go down different paths of design. In the last war, German Blitzkrieg tactics depended on fast-moving tanks swamping the enemy. The Leopard 1 is the direct descendant of this

thinking – it is a lightly armoured but highly mobile tank.

On the other hand, the British experience with light tanks was something of a disaster and they found that large, well-protected tanks were the most successful. Never again were British tanks going to be outgunned, and in its time the L11 120-mm gun on the Chieftain was the most powerful gun in the world.

Israel is always the exception. Having such a small population, it puts a very great emphasis on protection. The Merkava is the only

tank with the engine compartment at the front, thus adding to the protection for the driver.

More firepower

New technology has meant more and more penetrative power from existing gun designs. The British 120-mm L11 gun, the only 120-mm rifled gun, now fires the armour piercing, fin stabilised, discarding sabot round, known as 'fin'. Introduced in the early 1980s after observations on the effectiveness of other rounds in the Iran/Iraq war, fin will defeat all known

armour types.

The future of the tank is always open to speculation. Smart weapons attacking the thin top armour have meant a reappraisal of design philosophy. Equally, liquid propellant or even electromagnetic rail guns are being investigated to replace the existing guns. In the meantime work is under way on replacing the L11 with a new high-pressure gun with an even greater range and accuracy. The balance will constantly shift, but always around the three factors – firepower, protection and mobility.

1 Firepower

The standard NATO gun of the 1970s was the British-designed L7 105-mm single-piece ammo gun. It was extremely accurate, simple to load and produced under licence around the world. Unfortunately, British gun design tends towards rifled barrels, while the rest of the world has gone for smoothbore guns. The difference is fundamentally that smoothbore guns can fire rounds at a higher pressure and so have greater penetrative power. The world's most powerful gun is now the Rheinmetall 120-mm smoothbore, found on the German Leopard 2 and the US M1A2.

Both the Russians and Americans have experimented with firing missiles from tanks. The M551 Sheridan fired the Shillelagh down its 152-mm short barrel, while the T-64, T-72 and T-80 can all fire the AT-8 'Songster' missile with a reputed range of 4000 metres.

General Dynamics M1

After the demise of the joint US/UK/German tank venture known as the MBT-70, the US Army went on alone to produce a new tank to replace its fleet of M60s. The final design was approved in 1973 and designated XM1. It was to be a radical design right from the start, with a gas turbine engine replacing the usual diesel powerplant. The technical problems took time to solve, and the tank was not accepted into service by the US Army until 1982. From that point on, the Abrams became respected as perhaps the world's best tank, proving its potency in the Gulf War.

The tank is equipped with the latest technology available in tank design. The fire control computer is a Canadian design, and has proved so successful that it has been adopted by Vickers for the Challenger 2. It included a laser rangefinder linked into the stabilised sight: the gunner merely places the aiming mark on to the target and fires the laser. The computer will take in data such as range, air temperature and pressure, wind speed, ammunition type, barrel wear, ammunition temperature, and angle of tilt of the gun (known as trunnion tilt). The calculation is done in seconds and the computer will apply the correct elevation and aim off to the gun to ensure a first-round hit. The hardware includes a self diagnosing test facility that allows the gunner to check his equipment and locate any fault in seconds.

Further developments included the M1A1, with replacement of the 105-mm gun with the Rheinmetall 120-mm smoothbore, substantially improving the tank's firepower. Some M1s and M1A1s have depleted uranium (non-radioactive) based armour on the hull and turret; it is not clear if this will be retrofitted to earlier models. It is understood that this increases the weight of the tank from 57 tonnes to 65 tonnes.

Fume extractor

The swelling on the barrel about halfway down is a fume extractor, which evacuates the fumes caused by firing. As the round moves down the barrel it causes a vacuum behind it. When it passes the fume extractor, fresh air is drawn in behind it, forcing the fumes out of the muzzle as the round leaves. The only time it will not work is when the barrel faces directly into the wind.

Muzzle reference system

At the muzzle is the mirror of the Muzzle Reference System. When the gun fires, the force will cause a minute shift in the gun mounting. If this is not corrected, the relationship between gun and gunner's sight, and thus accuracy, is lost. The MRS works by projecting a light from a fixed source along the barrel off the mirror at the muzzle; the reflected beam is directed back into the gunner's sight. If the gun/sight relationship is OK, the spot of light will be shining in the centre of the aiming mark. If not, the gunner can realign his sight quickly and simply.

105-mm M68E1 gun

The M68 is in essence the same gun as the British L7. The first service example was on the Centurion Mk 5 in 1959, and since then it has become one of the most widely used guns in the world. The only real change is the breech block, which has been changed to accommodate a US-designed breech loading mechanism in which the breech moves vertically, as opposed to horizontally in the original L7.

Abrams

Turret crew

The tank and turret are commanded by the commander, who sits above and to the right. He controls all aspects of the tank, from selecting targets and controlling the gun through mapreading and giving instructions to the driver. The loader stands to the left. His main task is to ensure that the gun is kept loaded and ready to fire. The gunner sits below the commander and actually fires the gun.

Armour

Chobham armour was a British development designed to combat the threat from the new generation of improved anti-tank weapons. The exact composition is still classified. It obviously has a major steel component to provide the basic protection; laminated with this is a layer of the plastic Kevlar as found in body armour. This is a very good shock absorber. There is also a layer of ceramic material similar to that now found in high temperature engines, which will not melt except at extremely high temperature, so it is useful in defeating HEAT rounds. There is certainly one air space, which defeats HESH-type rounds. Chobham tanks are easy to identify, being slab-sided; the manufacturing process does not allow curved armour.

Driving compartment

The fourth crew member, the driver, sits on his own in a cab. The tank is steered by motorcycle-type handlebars; at each end of the T-bar are twist grip controls that work as throttles. Steering works by slowing one track and putting more power onto the other. It is possible to turn the tank on its own axis by driving one track forward and the other backwards.

Left: A rather rusty Chieftain barrel nevertheless gives a good idea of the rifling. In its time the L11 gun was the most powerful in the world, but it has now been surpassed by both the Rheinmetall 120-mm smoothbore found on the Leopard 2 and M1A2 and the GIAT 120-mm on the French Le Clerc and AMX-40. Britain is now the only country that persists in using the rifled barrel.

Far right: The gunner's station of the US M60. The gun equipment has self-testing circuits that alert the gunner to any faults in the system. It will identify the location of the fault, thus allowing quick repair.

Left: The new Challenger 2 is based heavily on Challenger, with the same Chobham armour and hull, but has an all-new turret. The box above the gun is part of the improved Thermal Observation and Gunnery System, which means the tank can operate equally well by day or night. The commander's cupola has been replaced by a French sight, giving the commander 360-degree vision. The gun is a high pressure variant of the L11, known as the XL30. It is still a 120-mm rifled bore but has an all-new loading and firing mechanism. Challenger 2 won in competition with the Leopard 2 and M1A2 to become the new British Army tank.

Below: The Soviet Army adopted 125-mm smoothbore as its standard calibre. The earlier tanks, T-55 and T-62, had 100-mm and 115-mm guns, which had inadequate firepower. The 125-mm fires fin-stabilised sabot and HEAT. The gun has some degree of stabilisation so it can be fired on the move, although it is less accurate than most Western tanks when doing so.

Above: The M551 Sheridan tank was something of a disaster due to the performance of its Shillelagh missile. This is not an outstanding weapon, but is reasonable enough. it was trialled on a variant of the M60 known as the M60A2, but was rejected. It is interesting that the less technologically advanced Soviets claimed to have developed a 125-mm anti-tank missile fired from a tank, seemingly with no problems at all.

Right: For most of the 1970s and 1980s the British-designed L7 105-mm rifled-bore gun was the NATO standard weapon – standard, that is, apart from Britain, which had the L11 120-mm gun. The 105-mm is still the most popular calibre, and the L7 gun or a variant of it has been fitted to a large number of tanks, including the M60 and M1. The L7 is a very accurate weapon: the 105-mm fin round is better than the 120-mm equivalent.

The radical Swedish S-tank has no turret: when it gets into position, with just the barrel and sights showing, it presents an incredibly small target. The gun is fixed – it is elevated and depressed by shifting the entire hull.

3 Mobility

The real measure of a tank's mobility is its bhp/tonne ratio: the ratio between the output of the engine and the weight of the tank. The higher the ratio, the more mobile the tank. Thus if we look at Leopard 1, Chieftain and T-62, they are 1.58, 1.04 and 1.10 respectively, Chieftain proving itself to be rather slow and cumbersome. If we look at the modern tanks like M1, Challenger and T-72, we find the respective figures are now 2.14, 1.5 and 1.45. But all this is slightly misleading since there are other factors to consider – Challenger, although having a lower power-to-weight ratio than Leopard 2, can actually beat it cross-country because it has superior suspension. Nothing is as simple as it seems!

A 1,500-bhp engine and a weight of 60 tonnes give the Leopard 2 an outstanding performance. On the road it will do 72km/h (45mph).

Different countries' experiences have taught them different things. By British standards, the French AMX-30 is woefully under-armoured. Equally, by French standards, British tanks are monstrously slow. The AMX has an impressive 65 km/h top speed.

Above: The fitting of ERA (explosive reactive armour) to Soviet tanks sent a shock wave through NATO. ERA will all but defeat HESH and seriously degrade HEAT; it has been seen fitted to the T-80 and T-64 and is being fitted to the T-72. ERA is lighter than other armours but is easier to defeat – it can be detonated by artillery and perhaps heavy machine-gun fire.

Left: Unique in tank design, the Israeli Merkava has the engine compartment at the front. This improves the protection factor, but at a terrible weight cost. Nevertheless, the Israeli experience has shown them that mobility is no substitute for protection, and the tank is designed around crew protection and escape.

NO STEP

Suspension

Suspension is important, not only for providing a comfortable ride for the crew, but more importantly a stable platform for the gun. The Abrams has torsion bar suspension in which each wheel is on its own mini-axle. This allows the tank to operate with the loss of some road wheels; it also gives 381 mm (15 in) of vertical travel. There are shock absorbers on the first, second and seventh road wheels, which allow the vehicle to travel at speed with the force of the bumps being taken up by the shock absorbers rather than the crew!

Turret bustle

The turret bustle has specially designed blow-off panels that, in the event of the ammo being hit, will explode outward, taking the blast away from the turret. Forty-four of the 55 rounds are stowed in this bustle, eight are in the hull, and three are carried externally in spallproof mountings in the turret baskets.

Turret

The commander is provided with six periscopes to provide 360-degree vision and a sight for the .50 Browning anti-aircraft gun. He is also linked to the gunner's primary sight. This has a magnifcation of ×10 (narrow field of view) or x3 (wide field of view). The sight is fully stabilised, so allowing accurate firing on the move. It has night vision capability in·the form of a passive thermal system. The crew are fed with filtered air which enables them to operate in a chemical and nuclear environment without the need for protective masks.

Lycoming Textron AGT 1500 engine

The gas turbine engine on the M1 was unique in the MBT world until the introduction of the T-80 by the Soviets. Although the gas turbine presents enormous technical problems, it points the way ahead. One of the aims in tank design is to minimise the weight: this can only be done by reducing the size of the components. One of the advantages of a gas turbine is that it is enormously efficient, about twice as efficient (weight for weight) than a conventional engine. This means twice as much power for the same weight. Another advantage of the gas turbine is that it is very quiet; with a conventional diesel the noise of the engine is deafening. The only noise you hear when the M1 approaches is the squeaking of the tracks – a very significant tactical advantage.

Secondary armament

The commander has a .50-cal (12.7-mm) anti-aircraft machine-gun. There are 1,000 rounds for this weapon. The loader has a 7.62-mm Belgian FN machine-gun, similar to that found on British tanks. Co-axially mounted with the M68 is an M240 7.62-mm (.30-in) machine-gun, also a variant of the Belgian FN MAG.

2 Protection

If you protect your tank with heavy armour, it slows it down. But if you don't, and it gets hit, you are in trouble. In the past, armour was made of solid steel. Modern armour has taken two directions. The British Chobham, a composite of various materials, is highly classified but thought to include Kevlar plastics and a ceramic layer. Chobham tanks are easy to identify, being slab-sided and angular. A variant applied to the Chieftain tank produced the Chieftain Stillbrew. Again, the material remains classified, but it was said to give the Chieftain the same protective power as the Leopard 2.

The Russians have opted for ERA (explosive reactive armour). Still secret, it is understood to comprise three layers of explosive designed to explode outwards when hit by a round, thus disrupting the incoming attack. Very effective against HESH and HEAT, it is of limited protection against fin rounds.

Above: The Continental AVDS-1790-2A 12-cylinder air-cooled diesel powerpack of the US M60 tank is situated behind rear-armoured doors that swing open and allow easy access to the engine and gearbox. They also allow very quick engine changes, unlike on British tanks.

Left: While the British were introducing the Chieftain, the Germans developed the Leopard 1. The two tanks reflect opposite thinking; the Germans went for lightly armoured, fast tanks and the British for slow, well-protected vehicles. German thinking prevailed; Australia, Canada and eight European nations bought it.

Below: The gas turbine engine of the M1 Abrams gives it an excellent power-to-weight ratio. But it is not without problems: it has a very high fuel consumption and requires non-standard fuel. Since it runs at 1,200 degrees it has a massive thermal signature that is easily spotted by detectors.

World War II saw the first real use of armour and quickly taught the Allies how superior the best of the German tanks were.

The fear of the German Panzer division inspired panic in the American forces. The German attack had been a complete surprise.

Tank vs Tank

Brigadier General McAuliffe, Deputy Commander, 101st Airborne Division, when called upon to surrender, replied in true American style: "Aw, nuts!"

"All I had to do was cross the River Meuse, capture Brussels, and then go on and take the port of Antwerp, and all this in December, January and February, the worst three months of the year in the Ardennes, where the snow is waist deep and there isn't room to deploy four tanks abreast, let alone six armoured divisions, when it didn't get light until eight in the morning and was dark again by four in the afternoon and my tanks can't fight at night, with divisions that had just been reformed and were composed chiefly of raw, untrained recruits – and all this at Christmas time!" – Sepp Deitrich, one-time Sergeant Major, later Commander, Sixth SS Panzer Army.

For a few glorious days, Dietrich's fears seemed unfounded, though the enterprise was doomed to failure once it left the close country of the Ardennes. But while cool analysis would have shown the Allies that that was the case, panic reigned instead, both on the battlefield and in the streets back home, while the Germans exulted.

"This time, we are a thousand times better off than you at home!" wrote a triumphant Leutnant Rockhammer in the early stages of the battle. "You cannot imagine what glorious hours and days we are experiencing now. It looks as if the Amis cannot withstand our push.

"Today we overtook a fleeing column and finished it off. We overtook it by taking a back road through the woods to the retreat line of the Ami vehicles then, just like on manoeuvres, we pulled up along the road with 60 Panthers. Then came the endless convoy driving in two columns, side by side, hubcap to hubcap, filled to the brim with soldiers. And then a concentrated fire from 60 cannon and 120 machine-guns . . ."

The American reaction to the Panzers' surprise attack was poor. The area from which the Germans launched their counter-attack was known as the Ghost Front, so quiet had it been through the autumn. But now, from north to south, the Germans hit five American units, four of them 'green' and the other battle-weary and understrength. Suddenly, there were white-smocked German paratroopers everywhere . . .

At the various headquarters up and down the front reports of exceedingly heavy shelling started coming in, each one more alarming than the last. Meanwhile, outside in the border villages, half-dressed, panic-stricken civilians were already grabbing their pathetic belongings and heading for the cellars.

The narrow, winding lanes leading from the front rapidly became impassable as vehicles and hordes of demoralised troops

The outstanding Königstiger or King Tiger. It outclassed anything that the Allies had, both in its gun and armour protection. Its only problem was its weight and fuel consumption, which restricted its deployment.

fought their way through slush and mud. Tanks and trucks skidded on the icy roads and inclines, often careering into the drainage ditches. Vehicles that ran out of fuel were simply abandoned; platoon leaders ordered their men to ditch their heavy equipment so that they could move faster.

Such was the fear that Germany's Panzer Divisions could inspire in men. As the historian of the 106th Division (which had only been in the line for five days) has written: "Let's get down to hard facts. Panic, sheer unreasoning panic, flamed that road all day and into the night. Everyone who had an excuse, and many who had

The US Army reckoned that it took five Shermans to knock out one PZKpfw V Panther. Typical of German tanks, the Panther was well armoured and had a big gun, the 75-mm KwK 42 L/70. It could penetrate 120mm of sloped plate armour at 1000 metres: this gave it a huge superiority over Allied tanks.

A US M4 Sherman of the 3rd Armoured Division passes a dead Tiger just outside Bovigny in Belgium. The M4 proved to be very useful if, by German reckoning, undergunned. It had either a 75-mm or later a 76-mm medium velocity gun.

none, was going west, away from Schoenberg and St Vith."

Germany may have been short of fuel and other strategic materials, and desperately short of trained men, but it certainly didn't lack effective tools. The 45-ton Panther was the best design of its generation, and then came the Tiger II. Weighing in at 68 tons, with a version of the celebrated 88-mm cannon that could pierce 200mm of homogenous armour at 1,000 yards, it was just about invincible.

Six months earlier, in Normandy, when the American, British and Canadian invaders were breaking out of the bridgehead, they had encountered Tigers in small numbers and were astounded at the amount of damage they could cause to the Allied armoured vehicles.

Twenty-five-year-old Leutnant Michel Wittman of the *Leibstandarte* SS Panzer Division spotted British tanks moving on a road some way below them. He was watching the lead Brigade of the British 7th Armoured Division, the 'Desert Rats'.

Wittman may have been brave, but he wasn't foolhardy. He waited as more and more tanks emerged from the dust cloud, very conscious of the imbalance of numbers; there were scores of them, and just one of him. In the end, it was an offhand remark of Wohl, the gunner, that stirred him into action. The British infantry were on both sides of the road, making no attempt to use cover. "Look at them!" said Wohl. "They're acting as if they've won the war already!" Wittman gave the order to move out, determined to shake the British out of their complacency.

The Tiger's 88-mm KwK-36 cannon opened up on the half-tracks carrying the infantry. Vehicle after vehicle exploded and burned, the troops inside pouring out and fleeing for their lives. Within five minutes, Wittmann's lone tank had knocked out 25 infantry carriers, and effectively blocked the advance of the British 22nd Brigade.

Brigadier Hinde, the British commander, managed to extricate his scout car from the general chaos, and started to organise his brigade while a dozen Cromwell tanks gave him what cover they could. But then four more Tigers arrived, and set about slaughtering the 8th Irish Hussars who manned them. Within minutes, all were either in ruins on the roadside or fleeing for their lives.

The noise of the battle attracted yet more German armour. Eight more appeared, under the command of SS Captain Moebius, and now it was only the slender resource of the Brigade's anti-tank group, with their poor six-pounder guns, who stood between the British and a stunning defeat. The Tiger's armour repelled the British anti-tank shells like ducks shaking off raindrops, and the only casualty was Wittmann's tank, hit in the track and brought to a halt. Wittmann ordered his men to bale out, happy to have brought the British advance to a complete standstill at the cost of just one tank.

In just one of these Tigers, Lieutenant Michel Wittman of the 12th SS Panzer Division destroyed 25 infantry carriers and held up the entire British 22nd Brigade. The Tiger suffered from being overweight and underpowered. Nevertheless, if used correctly, as it obviously was in Wittman's case, and if its rear armour was protected, there was little that could be done to harm it.

Above: The Panther was first used on the Eastern Front at Hitler's personal insistence. It was not ready for combat and suffered from terrible mechanical shortcomings, but by the time of the Normandy invasions these had been overcome and the tank was rightly feared by the Allies. In 1944 and 1945 nearly 4,000 Panthers were built, more than any other German tank during that time.

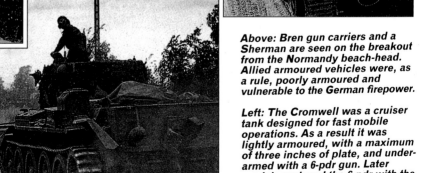

Above: Bren gun carriers and a Sherman are seen on the breakout from the Normandy beach-head. Allied armoured vehicles were, as a rule, poorly armoured and vulnerable to the German firepower.

Left: The Cromwell was a cruiser tank designed for fast mobile operations. As a result it was lightly armoured, with a maximum of three inches of plate, and under-armed with a 6-pdr gun. Later models replaced the 6-pdr with the same gun as the Sherman (75 mm).

TACTICS

FIN, TANK, ON

Firing on the move is one of the hardest skills to master. With the tank crashing around it is very difficult to keep the gun accurately on the target long enough to get a good shot in: there is a lot of luck in it.

The sudden appearance of an enemy tank in the sights springs the crew into a well-rehearsed drill; everyone knows what he has to do, and how to do it. Despite modern computers and sophisticated electronics, it is still down to the ability of the crew to operate as a single, highly-trained and professional team if they are to survive.

1 Scan the arcs

Once in position, the crew will scan their arcs; the commander will take the far distance and the gunner the close-range area. Both of them have identical gun control equipment, known as the gunner's or commander's grip switch and gun controller. By the use of a small joystick controlled by the thumb, the 17-tonne Challenger turret can be precisely traversed to sweep the arc. The commander's controls will, however, always override the gunner's. On sighting a target, cry "Target left" or "Target right".

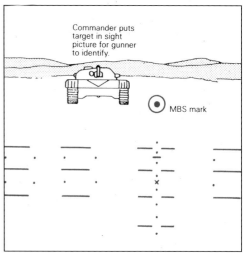

Commander puts target in sight picture for gunner to identify.

MBS mark

2 "Fin, tank, ON!"

Having sighted his target, the commander will lay the gun in the direction of the target so that the gunner can identify it. Both the commander's and gunner's sights are linked to the gun so that wherever it points their sights will follow. The commander will issue the fire order "Fin, tank, ON". 'Fin' tells the gunner and the loader which ammo type to use; 'tank' tells the gunner what to look for, and 'ON' tells him that the target is in sight and that he has passed over control of the gun.

TACTICS

A constant stream of coded messages comes into each ear, drowned by the shouts of your driver: "Which way now?" All this is accompanied by the constant growl of the diesel engine, the drone of the numerous cooling fans, filtration units and motors, the staccato barking of the machine-gun and the ear-splitting crash as the 120-mm L11 gun fires. You must make split-second decisions while being thrown around as you crash across the rough country. The turret points in one direction, the hull in another and you are looking in a third. Fighting in a modern tank is a highly skilled task that requires concentration and stamina. But how do you fight your tank against an enemy firing at you?

1 The commander sits in his own cupola. This is a mini-turret capable of independent rotation from the rest of the turret. He is provided with excellent sights for maximum observation: a typical sight has a magnification of ×10. It is his job to control the movement of the tank, map read, select ammo, select fire positions, acquire targets, encode and decode messages, and monitor two radio nets simultaneously.

Ammo selection is very important. The commander must try to ensure that he matches the type of ammo to the target. To destroy a tank, only AP rounds will do. Modern AP rounds are usually referred to as 'fin' or APFSDS (armour piercing, fin stabilised, discarding sabot). APCs and other lightly armoured targets are best engaged with either HEAT or HESH. It is important to select the correct ammo to avoid wastage – a Chieftain, for instance, carried only 52 rounds (32 fin, 20 HESH).

2 The loader/operator is responsible for loading the gun. He is also responsible for operating the radios (most tanks carry two sets, which require constant and simultaneous monitoring – one set in each ear). Operating radios was once a highly skilled task requiring constant attention, but modern communications devices are much simpler.

3 The gunner is the man who actually fires the weapon systems, either the main gun or the co-axially mounted machine-gun (known as the co-ax). He is also responsible for target acquisition as well as maintenance on all the gun-firing equipment.

Before the invention of passive night-vision devices, many tanks had powerful searchlights fitted. Many, like the one on this M60, had the ability to project either white light or infra-red, which is invisible to the human eye but was all too visible with detectors.

3 "Loaded!"

The gunner has spotted the target and correctly identified it. He will shout "ON", indicating to the commander that not only has he identified the target but also that he has control of the gun. If he does not identify the target, he will shout "Target not identified", in which case the commander will fine-lay the gun onto the exact target. At the same time, the loader will be loading the two-piece 120-mm ammo as fast as he can. Once the gun is loaded he will pick up the next round to be fired, ensure that the breech is closed and that the turret safety switch is on live, and shout "Loaded!"

Gunner fires laser to acquire range. Ballistic aiming mark electronically injected into sight picture.

Ballistic aiming mark

4 "Fire!"

The commander will now check that the gunner has laid on the right target and selected the right ammunition type on the computer, that the gun is ready to fire, and that the firing circuits are live. He will have supervised the loader to make sure that the correct round and charge were loaded. When he's satisfied he will shout the command "Fire!" This does not tell the gunner to fire the gun straight away, but to begin the firing sequence.

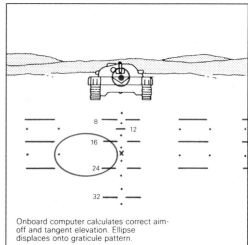

Onboard computer calculates correct aim-off and tangent elevation. Ellipse displaces onto graticule pattern.

5 "Lasing!"

The gunner will fire the laser to get the range to the target, shouting "Lasing!" as he does so. Less than a second later the ballistic aiming mark (an ellipse) will be electronically injected into the sight picture. The computer will calculate the correct elevation and aim off for the gun. The ballistic aiming mark will shift from around the target onto the engraved graticule pattern. A second later, the gun will automatically drive up so that the ellipse is once again around the target. When it is, the gunner will fire, shouting "Firing now!"

4 Although modern tanks are highly protected, the correct use of ground for protection is vital. Tanks need to get into fire positions in the same way as the infantry do. In essence, a tank fire position will be on the back of a slope and the driver will position the tank so that only the turret is ever exposed above the crest. The formula used is CRAB MEAL (Cover, Routes, Arcs, Background, Mutual support, Enemy, Air, Landmarks).

He must try to avoid obvious features that are out in the open. Remember the enemy direction, and ensure that your main killing arc faces it. Think of routes in and out, and don't put your tank against an obvious background.

At the receiving end. There are three categories of tank kill. An M-kill is a mobility kill: the tank cannot move but otherwise is fully functional. In a P-kill the crew is killed but the tank survives. In a a K-kill, neither crew nor vehicle are functional.

Left: German Leopard Is fire at night. The muzzle flash, tracer burn and target strike can be seen. Most rounds have tracers in the base to indicate its flight to the crew and so enable the gunner to make adjustments if he misses.

Right: The Israeli Armoured Corps is the world's most experienced tank force. Western armies have based a lot of their tactics on Israel's experiences, but fighting Arab armies is not quite the same threat that Western armies may face.

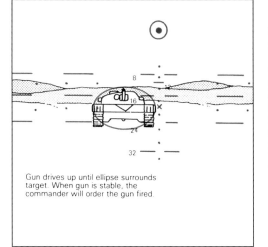

Gun drives up until ellipse surrounds target. When gun is stable, the commander will order the gun fired.

6 "Target!"

There will be a huge explosion and a blinding flash as the gun fires. The tank will rock slightly. Both gunner and commander observe the fall of shot. The chances are that, if the equipment is functioning correctly and the proper drills were observed, the round will hit the target. A direct hit using a fin round will result in a blinding white flash. The gunner will shout "Target!", indicating a hit. The commander will reply "Target stop", confirming the hit and finishing the engagement sequence. The loader will still load the next round to ensure that the gun is always ready.

7 "Target stop"

If the preceding engagement sequences have been followed properly and the equipment is functioning correctly, then there is an almost 100 per cent chance that the round will hit the target. That is not to say that the target will be destroyed: if, for instance, the target was moving, the first hit may give an M-kill. It will then be necessary to shoot it again to achieve K-kill. Experience from modern wars, and analysis of the ground over which the next war may be fought, has shown that most tank engagements are likely to take place at around 1000 metres. But current British tanks are designed to engage at over 3200 metres and to

start engaging at that range is something of a waste of ammunition since the chances of hitting are greatly reduced. Also, at very close ranges (less than 400 metres) there are special techniques for engaging targets. At that range all you have to do is point the gun at the thing and fire: you are going to hit something. When planning a tank shoot, an ideal position is one in which the commander can observe the targets at longer ranges, for instance 3000 metres, and get a good fix on them. When they come into battle range of around 1500 metres he can then engage with an almost certain chance of hitting.

US troops fan out in a defensive pattern from a GM LAV-25, a Light Armoured Vehicle initially designed to meet a USMC requirement. It is armed with a 25-mm cannon, a 7.62-mm machine gun and eight smoke-grenade dischargers.

MECHANISED INFANTRY

Are they battlefield taxis for carting grunts around, or combat vehicles designed to fight beside the tanks?

The Israeli army is the largest user of the US M113 after the US Army. They have tried it in combat on numerous occasions, and it has proved itself over and over again. Although now outmoded by more sophisticated and better-armed vehicles, the M113 remains the definitive APC of its time.

"Move and see! Move and fight! Suppress and destroy!" screams the United States Army's operations manual for the Mechanised Infantry Battalion. Those simple phrases sum up the whole aggressive philosophy of modern warfare today, just as they did 60 years ago, when Guderian, the inventor of the Blitzkreig, became one of the most feared men in the world.

The truck, the half-tracked weapons carrier, the tank and the self-propelled gun changed the face of war for ever when they were first used as a single unit in the lightning attack that overran France and the Low Countries in 1940. But even as late as 1944, when the Panzer-SS Das Reich Division was hurrying from its base at Montauban to try to contain the Allied beach-head in Normandy, many of its support vehicles, even its anti-aircraft guns, were still horse-drawn.

The Panzers were unstoppable in 1940 for a simple reason – speed and aggression of purpose. They were unstoppable on the Eastern Front a year later too, until the Soviets' best ally, winter, took a hand in the game. The Russian winter had previously beaten better generals than Guderian.

The underlying principle is simple: speed and mobility wins battles, even against superior odds. Yet every commander knows that mobility leads to a loss of communications, and that inevitably leads to chaos. To be successful the infantry has to be more than just mounted; it has to be provided with an infallible communications system, to allow the commander to direct its operations, and it must be equipped with weapons that are up to the job at hand.

"Defence is temporary; awaiting opportunity for offense," says US Army doctrine, perhaps rather optimistically. The manual then goes on to describe the way "the Threat" is likely to organise his offensive: his "concentration of numerically-superior forces and firepower for a combination of frontal attacks, enveloping manoeuvres, and deep offensive thrusts into your rear by armour-heavy combined arms forces." This, it suggests, will allow him to advance between 30-50 kilometres (20-30 miles) a day.

Flexibility

It doesn't sound too much. A couple of hours on a bicycle? A day's hiking? A quarter of an hour in a helicopter? But the reality is rather different. An advance like that, certainly if it had continued for days, would have been called a rout in World War II. It's the sort of advance that the North Koreans made in 1950, which was certainly described in those terms. But if that sort of flexibility is commonplace today, then it has vital lessons for defenders, the first of which is: static defence won't work unless it denies the enemy his mobility.

Just as the enemy will be constantly probing, seeking a weak spot to insert his attack-

ing forces and punch through, so the defender must use the mobility that mechanisation gives him. Combined with artillery and fixed- and rotary-wing air strikes, the defender must slow him down and stop him; hopefully in a position where he's vulnerable to counter-attack, creating the "opportunity for offense"

Armed with a 20-mm cannon and a 7.62-mm co-axial machine-gun, the French AMX-10P is in line with current designs of infantry fighting vehicles.

Desert War

FLASHBACK

Of all the countries of the world, perhaps the one with the most experience of fighting the highly mobile war of the mechanised infantry is Israel. During the fight for their very existence back in 1948, the Haganah, the forerunner of the Israel Defence Force, ran convoys of armoured trucks through the gauntlet of Arab forces to supply the besieged city of Jerusalem.

The Israeli basic vehicle was, for a long time, the rather old US Army M2 halftrack. The vehicle dates from World War II, but once in Israeli hands it was quickly refurbished and formed the backbone of the IDF during the campaigns of 1956, 1967 and 1973. It is still in service today, used as a command post, ambulance or anti-aircraft mounting. Nowadays, however, the IDF is the world's second-largest user of the US M113, which is known to the Israeli soldier as 'Zelda'.

The Israeli army, with probably more experience of mechanised infantry than any nation in the world, shuns the high-tech approach and goes for simple, cheap vehicles.

MECHANISED INFANTRY Reference File

181

USA

FMC Bradley

The US Army formulated a requirement for a mechanised infantry combat vehicle in the early 1960s. After two abortive starts, the XM723 that resulted from a 1972 specification led to the closely related **M2 Bradley Infantry Fighting Vehicle** and the **M3 Bradley Cavalry Fighting Vehicle**. These entered service in 1981 with the primary task of scouting for and supporting main battle tanks, the M3 having two rather than seven infantrymen to allow the carriage of 12 rather than seven TOW anti-tank missiles.

The vehicles are based on a welded aluminium armour hull with spaced laminate and applique steel armour in some areas. In the centre is the

electrically powered aluminium and steel turret armed with a stabilised M242 Bushmaster dual-feed cannon, a co-axial machine-gun, a retractable two-round missile launcher, and an advanced fire-control system. Standard features on these fully amphibious vehicles include an NBC protection system and night vision equipment. Several significantly improved versions of the Bradley have appeared over recent years, some with explosive reactive armour.

Specification
FMC M2 Bradley
Type: mechanised infantry fighting vehicle with a crew of three and provision for seven troops
Weight: 22.59 tonnes (22.23 tons)

Armament: one 25-mm cannon with 900 rounds, one 7.62-mm (0.3-in) co-axial machine-gun with 2,340 rounds, one twin-tube TOW missile launcher with seven missiles, and four smoke-dischargers on each side of the turret
Performance: maximum speed 66 km/h (41 mph); range 483 km (300

miles)
Dimensions: length overall 6.453 m (21 ft 2 in); width 3.2 m (10 ft 6 in)
Powerplant: one 373-kW (500-hp) Cummins VTA-503T diesel engine
User: USA and Saudi Arabia

Designed to complement their range of C1 and Leopard 1 main battle tanks, the Italian army has recently accepted the VCC-80 for trials. It is of typical early 1980s design, with cannon, co-axial machine-gun and high-speed engine.

The French AMX-VCI has been replaced by the AMX-10P. The vehicle's hull is based on that of the AMX-13 light tank.

Threat Principles

'THE THREAT' – US Army jargon for the forces of the Warsaw Pact - adhered to certain principles to enhance its doctrine of high-speed offensive action. To defeat the threat, you had to:

- **Seek surprise** at all times to paralyse the enemy's will to resist and deprive him of the ability to react effectively.
- **Achieve mass** in decisive areas by rapidly concentrating men, material and firepower for the minimum time necessary to rupture the enemy's defences.
- **Achieve flank security** by aggressive advance.
- **Breach enemy defences** at weakly defended positions and rapidly advance deep into rear areas.
- **Bypass strongly defended areas,** leaving them for neutralisation by following echelons.
- **Launch massive artillery support** to include mortars, multiple rocket-launchers, anti-tank guns and tanks.
- **Dedicate a high priority** to the enemy's nuclear and anti-tank weapon systems.
- **Employ tactical air support** to achieve air superiority and conduct air strikes on the enemy's rear areas.
- **Employ radio-electronics** (jamming systems) as a primary element of combat power.
- **Accept heavy losses** and the isolation of assault units.
- **Conduct operations 24 hours a day.**

182
GERMANY

Thyssen Henschel/Krupp MaK Marder

In the late 1950s West Germany developed a tracked chassis suitable for several military applications, and this was adopted for the **Marder** (marten) to replace the Swiss-designed Spz 12-3 used by the German army since its re-formation in the earlier 1950s.

At the time of its introduction in 1971, the Marder was the West's most advanced mechanised infantry combat vehicle, and it remains a formidable machine. The type is based on a welded steel hull with the driver and engine at the front left and right respectively, the two-man steel turret in the centre and the troop compartment at the rear, accessed by a power-operated rear ramp and surmounted by a remotely controlled

machine-gun. The electro-hydraulically powered main turret has a dual-feed MK 20 Rh 202 cannon and has been retrofitted with a MILAN anti-tank missile launcher. Upgraded vehicles are the **Marder A1** with cannon, night vision and habitability improvements reducing infantry accommodation from six to five, the **Marder A1A** with cannon and habitability improvements, and the **Marder 25** with a 25-mm Mauser Model E cannon.

Specification
Thyssen Henschel/Krupp MaK Schutzenpanzer Neu M-1966 Marder
Type: mechanised infantry combat vehicle with a crew of four and

provision for six troops
Weight: 28.2 tonnes (27.75 tons)
Armament: one 20-mm cannon with 1,250 rounds, two 7.62-mm (0.3-in) machine-guns (one co-axial and one over the rear hull) with 5,000 rounds, one MILAN anti-tank missile launcher
Performance: maximum speed

75 km/h (46.5 mph); range 520 km (323 miles)
Dimensions: length overall 6.79 m (22 ft 3 in); width 3.24 m (10 ft 8 in)
Powerplant: one 447-kW (600-hp) MTU MB 833 Ea 500 diesel engine
User: Germany

Warrior Crew Stations

The latest addition to the British Army is the Warrior Mechanised Infantry Fighting Vehicle, armed with the 30-mm RARDEN cannon with a co-axial 7.62-mm Chain Gun. There is no provision for the troops to fire their own weapons from inside the compartment, although this has been incorporated in the export variant. In order to crew the vehicle one man has been taken out of the fire teams and is now the gunner. Tactically, there is much work to be done on the best use of the Warrior vehicle once the troops have left it. Does it go to one side and shoot the squad in? Do you group all your empty vehicles together to form a platoon firebase? With the Warrior, it seems it was rather a case of inventing the weapon and then deciding what to do with it afterwards.

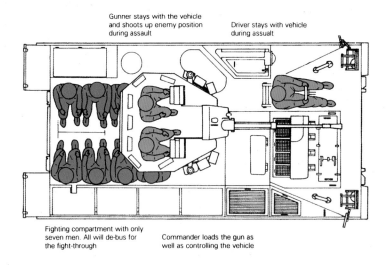

Gunner stays with the vehicle and shoots up enemy position during assault

Driver stays with vehicle during assualt

Fighting compartment with only seven men. All will de-bus for the fight-through

Commander loads the gun as well as controlling the vehicle

183

FRANCE

GIAT AMX-10P

Designed from the mid-1960s by the Atelier de Construction d'Issy-les-Moulineaux as replacement for the AMX VCI, the **AMX-10P** entered production at the Atelier de Construction Roanne in 1972 and entered service in 1973.

The AMX-10P is a potent mechanised infantry combat vehicle, and is based on a hull of welded aluminium armour construction with the driver and engine at the front left and right respectively, the turret in the centre, and the troop compartment in the rear. The electrically powered aluminium armour turret accommodates an M693 (F2) dual-feed cannon able to change instantly between HE and armour-piercing ammunition. The embarked

troops enter and leave the vehicle via a full-width electrically powered rear ramp. There is a whole range of derivatives for other battlefield tasks, though the only two MICV versions are the **AMX-10P Marine** for the Indonesian marines with improved waterjets and a rear-mounted heavy machine-gun turret increasing troop capacity to 13, and the **AMX-10P 25** with a more powerful 25-mm GIAT 811 dual-feed cannon.

Specification
GIAT AMX-10P
Type: mechanised infantry combat vehicle with a crew of three and provision for eight troops
Weight: 14.2 tonnes (13.975 tons)

Armament: one 20-mm cannon with 760 rounds, one 7.62-mm (0.3-in) co-axial machine-gun with 2,000 rounds, and two smoke-dischargers on each side of the turret
Performance: maximum speed 65 km/h (40.5 mph); range 600 km (373 miles)

Dimensions: length overall 5.778 m (18 ft 11 in); width 2.78 m (9 ft 1 in)
Powerplant: one 209-kW (280-hp) Renault (Hispano-Suiza) HS 115-2 diesel engine
Users: France, Greece, Indonesia, Qatar, Saudi Arabia and United Arab Emirates

---

184

ITALY

OTO Melara/IVECO VCC-80

After considerable preparatory work, the consortium formed by OTO Melara and IVECO (a Fiat subsidiary) began definitive work on a new mechanised infantry fighting vehicle in 1982, and this **VCC-80** appeared in prototype form in 1985. Throughout the design procedure emphasis was placed on high levels of firepower, mobility and protection, the last being aided by the vehicle's low silhouette.

The vehicle is based on a welded hull of aluminium armour construction with a bolt-on layer of steel ballistic armour. The driver and engine are located at the front left and right respectively. In the centre of the vehicle is the turret, an electrically powered unit of welded aluminium armour construction, and

this accommodates the stabilised Oerlikon-Bührle KBA-B02 cannon, which is a powerful dual-feed weapon firing HE and armour-piercing ammunition, and operated in conjunction with an advanced fire-control system including a laser rangefinder and thermal imaging night sights. At the rear is the troop compartment, which is accessed by a hydraulically operated rear ramp.

Specification
OTO Melara/IVECO VCC-80
Type: mechanised infantry combat vehicle with a crew of three and provision for six troops
Weight: about 19.00 tonnes (18.70 tons)

Armament: one 25-mm cannon and two 7.62-mm (0.3-in) machine-guns (one co-axial and one AA) each with an unrevealed number of rounds, and three smoke-dischargers on each side of the turret
Performance: maximum speed 70 km/h (43.5 mph); range 600 km (373

miles)
Dimensions: length overall 6.705 m (22 ft 0 in); width 2.98 m (9 ft 9 in)
Powerplant: one 358-kW (480-hp) Fiat diesel engine
User: currently entering production with the Italian Army

the doctrine describes.

Put this way, the modern battlefield starts to sound more like a boxing ring, and that's not a bad comparison. You can even take it one step further, and liken a clash between, say, unsupported companies, to a flyweight fight – fast and furious, with both sides relying on aggression, keeping nothing back for a long-drawn-out engagement. Battalion on battalion is more like a welterweight bout. There's still a lot of aggression, of coming to meet each other, but there's some tactical

nursing of strengths too, while looking for an opportunity to put in an overwhelming attack. And then we come to the heavyweights, when two fully equipped divisions meet.

Let us forget the first round knock-out, and imagine a more even contest. First, patrol and reconnaissance activity tries to probe for a weakness. If it finds one (and if it can't, then it must create one as best it may), the commander's task is to get a strong attack launched in that area before the defender can muster his forces. He will only have a limited

time in which to attack at long range, with artillery and aircraft; soon, his own ground forces, both armoured and infantry, will be engaging the enemy. First at medium range, with tank main armament and ATGWs, and then close-in.

With rapid deployment as top of the list of priorities, you have to examine each section of your force with its mobility in mind. Take the most versatile. That becomes the standard by which you have to measure all the others.

Left: The crew compartment of the Warrior is far from roomy despite the external size of the vehicle. The hull provides the troops with bullet- and splinter-proof protection as well as filtered air for operating in NBC conditions. On the more homely side, the vehicle has two BVs, or boiling vessels, which means that the section can cook its food centrally.

Right: What it's all about – debussing on to a position. With eight in the back, hot, dusty and extremely cramped, the chance to get out is a relief. Debussing drills must be very slick as the troops in the back have no idea where they are or where the enemy is when they pile out of the back. It is very disorientating.

185

SWITZERLAND

MOWAG Tornado MICV

The Mowag company has been engaged in the design and development of tracked and wheeled vehicles since just after World War II. In the 1970s, having secured the contract for the vehicle that later became the Marder, the company developed, as a private venture, the **Tornado**. The similarity between the vehicles is evident. At this time the standard APC of the Swiss army was the M113. With its decision to upgrade its tank force with Leopard 2, there was clearly a need to do likewise with its APCs. Tornado was seen as a contender.

The hull of the Tornado is of all-steel construction, with its sides and front well sloped for maximum protection. The driver is at the left front of the

vehicle, with the commander to his rear and the engine and transmission to his right. In the centre of the hull can be mounted a wide range of armament installations depending on the mission requirements. One of the most powerful is the Swiss Oerlikon-Bührle two-man power-operated Type GDD-AOE turret, which has an externally mounted 35-mm KDE cannon fed from two ready-use magazines.

Specification
MOWAG Tornado MICV
Type: mechanised infantry combat vehicle with a crew of three and provision for eight troops
Weight: 22.3 tonnes (21.95 tons)
Armament: one 35-mm cannon with

100 rounds and 7.62-mm co-axial machine-gun with 500 rounds; two externally mounted 7.62-mm machine-guns at the rear with 500 rounds apiece
Performance: maximum speed 66 km/h (41 mph); range 400 km (249 miles)
Dimensions: length overall 6.7 m

(22 ft); width 3.15 m (10 ft 4 in); height 1.75 m (5 ft 9 in)
Powerplant: one Detroit Diesel Model 8V- 71T developing 290 kW (390 bhp)
Users: No longer in development

186

USA

FMC Corporation M113

The definitive APC of its time has to be the **M113**. It has proved to be the most successful APC in the free world and is in service with almost every NATO nation, with the glaring exception of the UK. The FV432 is very heavily based on the M113, although inferior.

The M113 was introduced into service in the US Army in May 1963 after trials between a steel and aluminium vehicle; the latter was selected. The engine compartment is to the right of the driver's position. Immediately behind the driver and to his right is the commander; in some variants he has his own cupola but on the basic vehicle he has only a hatch and vision blocks. The commander usually mans the Browning .50-cal.

The M113 has spawned an entire family of variants: too many to list in detail. However, the most numerous are the M577 command vehicle with raised roof to accommodate radio equipment and the map boards; the M125 81-mm mortar carrier; the M163 Vulcan anti-aircraft gun, and the M548 TLC, or tracked load-carrier.

The M113 is being phased out in the US Army to be replaced by the M2 Bradley but it will remain an amazingly popular vehicle for a long time to come.

Specification
FMC M113A2
Type: armoured personnel carrier with a crew of two and provision for 11 troops

Weight: 11.34 tonnes (11.16 tons)
Armament: depending on variant but typically one 0.5-in (12.7-mm) Browning heavy machine-gun
Performance: maximum road speed 67.59 km/h (42 mph); range 483 km (300 miles)
Dimensions: length 2.686 m (8 ft

9 in); width 2.54 m (8 ft 4 in); height 1.85 m (6 ft 1 in)
Powerplant: one GMC Detroit Diesel Model 6V53 developing 160 kW (215 bhp)
Users: 49 countries worldwide

The German Marder is an outstanding vehicle, armed with a 25-mm cannon and with two other 7.62-mm guns. The rear is protected by a remotely operated 7.62-mm gun that can cover troops as they remount via the rear ramp.

Tanks and APC-mounted infantry come top of the table. The tank's high speed across country, armour protection and firepower allow it to kill soft targets and other tanks at relatively long ranges and still stay invulnerable to small-arms and artillery. On the down side, it is at risk from infantry-operated anti-tank guided missiles, as well as the longer-established anti-tank gun and mine, and it doesn't see too well when it's forced to button up and close all hatches.

Fast and safe

The APC, on the other hand, gives the infantry the best of both worlds – or at least, something other than the worst, which is what they've traditionally had to bear. Invulnerable to small-arms fire, it can get troops to where they're needed, fast and in comparative safety. Its 50-calibre machine-guns (or larger-calibre cannon, in some cases) give

Combat Comparison

Recent generations of fighting vehicles are rather similar; even British and Russian ones seem to have converged.

187

FORMER USSR

BMP-2

The BMP-2 is the updated version of the BMP-1 mechanised infantry combat vehicle. It is not a replacement for it, but rather a supplement to it. It was first seen during the November 1982 Moscow parade, but reports of a new vehicle had been circulating some years before that. It has been in service with the Western Group of Forces (the permanent garrison in East Germany) for some years and saw operational use in Afghanistan.

The chassis of the BMP-2 is almost identical to its predecessor. The driver sits at the front of the left of the vehicle and is provided with three vision blocks for

operating closed down. The commander has moved to the turret, a change from BMP-1 where he is immediately behind the driver.

The vehicle is armed with a 30-mm RARDEN-type cannon and a co-axial PKT 7.62-mm machine-gun. The rather ancient AT-3 'Sagger' of the BMP-1 has been replaced by a mount for either AT-4 'Spigot' or AT-5 'Spandrel'; both of them a marked improvement on AT-3. The gun carries 500 rounds of ammo and has a maximum elevation of 74 degrees – clearly intended to be used either in an anti-helicopter role, or perhaps in built up areas. It is claimed to be effective against ground targets at up to 1000 metres.

Specification
BMP-2
Type: mechanised infantry combat vehicle with a crew of three and provision for six troops
Weight: 14.6 tonnes (14.37 tons)
Armament: one 30-mm cannon with 500 rounds, one AT-4 or AT-5 launcher with unknown number of missiles, one PKT 7.62-mm (0.30 in) co-axial machine-gun with 2000 rounds
Performance: maximum speed 80 km/h (50 mph); range 500 km (310 miles)
Dimensions: length 6.86 m (22 ft 6 in); width 3.09 m (10 ft 2 in); height 2.08 m (6 ft 10 in)
Powerplant: 224-kW (300 bhp) type UTD-20 6 cylinder inline, water-cooled diesel
Users: 28 countries worldwide

One of the most important factors to note between Warrior and BMP-2 is the difference in height of the vehicles. BMP-2 is very much lower, thus presenting a very much smaller target. It also has a long, sloping front, which means it is even less of a target. To its detriment, it has shockingly poor armour.

The turret is fitted with a 30-mm RARDEN-type cannon very similar to that of the Warrior. In addition, it has the anti-tank capability of the AT-4/5 missile.

The crew compartment of BMP-2 is very crowded and massively uncomfortable. The rear doors are hollow and serve as fuel tanks! An encouraging thought to go into battle with. Unlike Warrior, troops can fire their weapons under armour.

it an impressive ability to kill soft targets, even on the move, and if it mounts a TOW anti-tank missile then even heavy armour has to look out.

So what of the other components of the Mechanised Infantry Battalion – the heavy weapons, the mortars, the ATGW teams? Unless it can be given the same mobility as the rest of the combat team, its integrity will be lost. There is no alternative to mechanisation. All elements must be mounted and able to move with the same speed and decisiveness. This includes the all-important communications sections and whatever support arms – engineers, for example – that are attached.

Because the entire team must work as one, and there is so little time to lose, the commander becomes an even more important figure than previously. When infantry moved at its own pace – a few miles an hour at best, a few yards an hour more commonly – there was time for a commander to consider his actions comfortably, and to correct mistakes as and when they became apparent. In the cli-

mate of the modern, mechanised battlefield, moments are all he will have; an incident will no sooner blow up than it will be over, won or lost, probably with no second chance to recover a loss.

NATO armies recognise this factor, and accept that the battalion's ability is closely matched to that of its commander to com-

municate and motivate. The Warsaw Pact forces, especially the Soviet army, have long depended on the commander and a small cadre, and have never relied on individual acts of intelligence to see it through. When this doctrine works in practice, it is very formidable indeed, producing a concentrated, unified force, capable of moving and fighting as one.

The US M2/M3 Bradley replaces the M113. Note the size of the vehicle compared to the driver.

Warrior and BMP-2 represent modern designs for both countries. How do they compare?

188

GB

GKN MCV-80 Warrior

In service since 1987, the **MCV-80 Warrior** was designed to supplement rather than replace the FV432 tracked armoured personnel carrier, and is the UK's first true mechanised infantry combat vehicle.

The type was designed from 1972, and is based on a hull of welded aluminium armour construction with the driver and engine at the front left and right respectively, the welded-steel Vickers turret in the centre, and the troop compartment at the rear. The two-man turret accommodates the well-established L21 RARDEN cannon, a clip-loaded weapon firing HE or armour-piercing ammunition, and a licence-built Hughes EX-34 Chain Gun co-axial machine-gun. The embarked troops

enter and leave the vehicle via a power-operated door in the rear, but have no provision to fire their weapons from inside the vehicle; there are, however, two roof hatches. Standard features includes an NBC protection system, a manually operated fire-extinguishing system, and night vision equipment for the three-man crew. Several other models of the Warrior have been developed (Warrior Platoon, Command, Repair and Recovery, Engineer, and Artillery Observation Vehicles) including a fire-support model with a 90-mm gun in a powered turret.

The Perkins diesel engine gives the vehicle excellent mobility despite its 24 tonnes. The suspension gives the troops a far smoother ride than the BMP-2, and so they are in a better state when it comes to fighting.

The turret has an almost identical gun to BMP-2, but lacks any anti-tank capacity. It does, however, have superior day and night sights.

Specification
GKN Defence MCV-80 Warrior
Type: mechanised infantry combat vehicle with a crew of three and provision for eight troops
Weight: 24.5 tonnes (24.11 tons)
Armament: one 30-mm cannon and one 7.62-mm (0.3-in) co-axial machine-gun each with an unrevealed number of rounds, and four smoke-dischargers on each side of the turret
Performance: maximum speed 75 km/h (47 mph); range 500 km (311 miles)
Dimensions: length overall 6.34 m (20 ft 10 in); width 3.034 m (9 ft 11 in)
Powerplant: one 410-kW (550-hp) Perkins (Rolls-Royce) CV8 TCA diesel engine
Users: UK and Kuwait

The British Army puts greater emphasis on creature comforts than does the conscript Russian Army. This is reflected in the much larger troop compartment and basic amenities provided – such as two cookers for the troops' food.

RED ADVANCE

Soviet infantry debusses from the BMP-1. This was the world's first mechanised infantry combat vehicle.

With highly integrated organisations and thorough well-thought-out drills, the Russian Army is an imposing force to tackle. It quickly recognised the importance of proper balanced forces, and has developed its tactics to match.

The Russian Army is organised into the two types of ground combat arms: motor rifle troops, and armoured troops. These roughly equate to infantry and cavalry in Western forces, but this obscures the most important feature and strength of the Soviet organisation. Both types of troops are completely integrated: every motor rifle organisation has tanks in it, every tank structure has a motor rifle component. They never operate on their own.

In terms of operational units, the Russian Army is organised in roughly similar lines to Western forces.

The basic unit is a battalion, made up of 31 vehicles: either tanks or BMPs, depending on whether it is a tank or motor rifle battalion. These battalions are grouped within regiments: a Tank Regiment is made up of three tank battalions and one motor rifle battalion. A motor rifle regiment is the exact opposite.

Tank strength

Regiments are grouped within divisions; again three tank regiments and one motor rifle regiment or vice versa, depending on whether it is a tank or motor rifle unit. Furthermore, the Soviets are

BMP armoured personnel carrier

The BMP is the Rusian Army's main APC. It is designed, as is all Russian equipment, to be cheap to produce, reliable and simple to operate. It allows the infantryman to fight from under armoured protection and gives the motor rifle troops their own, albeit old, anti-tank missile: the AT-3 'Sagger' is mounted on every BMP-1. The BMP is seen in numerous variants, from a basic troop carrier right up to a command vehicle, recce vehicle, artillery locating radar vehicle and numerous others.

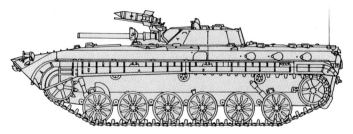

Armed with a 73-mm gun and AT-3 missile, the BMP is well equipped.

The Russian Army puts great emphasis on low-profile vehicles, which gives them a tactical advantage since they are harder to hit. The disadvantages are a dire lack of crew comfort and very thin frontal armour.

great believers in large tank forces, so in tank-weak organisations they add extra tanks. In a regiment there will be an independent tank company (10 tanks); in a division and an independent tank battalion (40 tanks, not 31 as in a normal tank battalion).

The combined arms philosophy of the Russian Army is very important. It rarely operates tanks without infantry, and vice-versa. Here the BMP-1 is fighting with the rather old T-55. The tank has been withdrawn from front-line forces, but the BMP remains.

Regiment in the advance

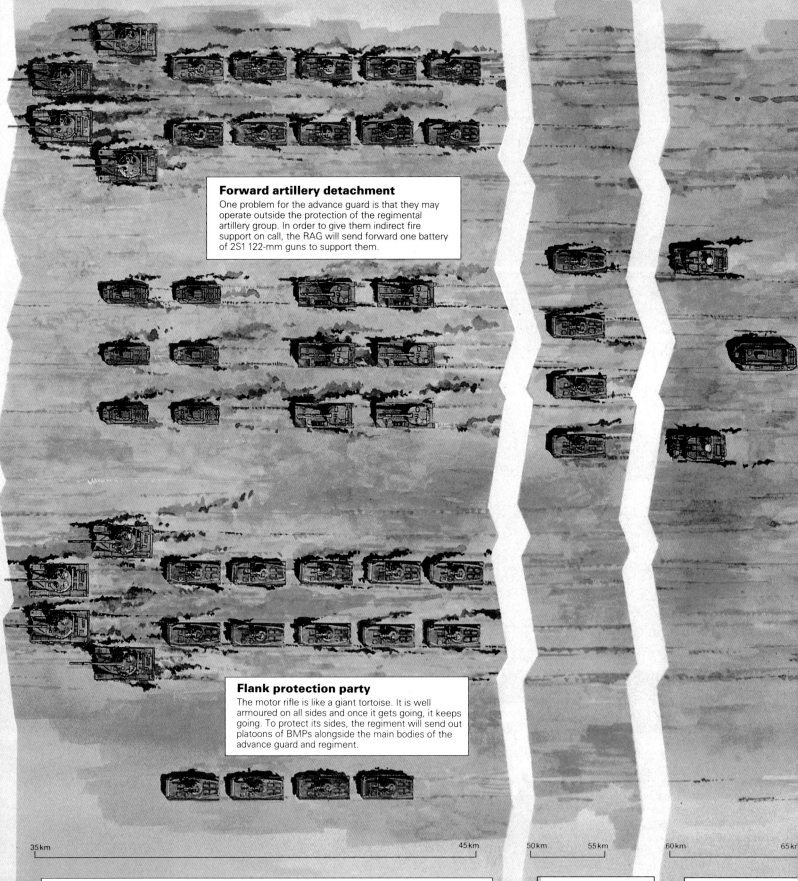

Forward artillery detachment
One problem for the advance guard is that they may operate outside the protection of the regimental artillery group. In order to give them indirect fire support on call, the RAG will send forward one battery of 2S1 122-mm guns to support them.

Flank protection party
The motor rifle is like a giant tortoise. It is well armoured on all sides and once it gets going, it keeps going. To protect its sides, the regiment will send out platoons of BMPs alongside the main bodies of the advance guard and regiment.

35 km   45 km   50 km   55 km   60 km   65 kr

The main body of the advance guard
The bulk of the advance guard travels in one large packet about 10 km long. It is a battalion of BMPs with additional tanks, artillery and anti-tank assets. It will operate as an entity, and its task is to deal with small pockets of enemy forces so that the rest of the regiment can continue without deploying from the line of march. The entire philosophy of the organisation of the regimental advance is that each level is a copy of the one above it. A regiment sends forward a battalion as its advance guard. The advance guard will need its own protection, which is why it sends forward the vanguard. They in turn send forward the CRPs. At no stage will any force be on its own or have nobody in front providing early warning. The whole emphasis of the advance is the maintenance of momentum.

Rear security patrol
To provide rear protection, a platoon of BMP-1s will be detailed off to travel some 5 km behind the main body. Their task is not to fight but to report and screen.

Regimental command group
The regimental commander will travel his own group, with air defence from the ZSU-23/4s. He may we be up with the advance guard where he can control operations.

Soviet Motor Rifle

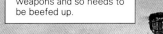

Mobile anti-tank patrol

Because the bulk of a motor rifle regiment is made up of BMPs, it is weak in anti-tank weapons and so needs to be beefed up.

0 km   5 km   15 km   20 km   25 km   30 km

Regimental reconnaissance

Leading the regimental advance will be the recce detachments, operating about 10 km in front of the CRPs. These are patrols of BRDM-2s and BMP-1s. The patrols will not attack since they lack the firepower, but they will be trying to establish locations and sizes of enemy forces.

Combat reconnaissance patrols

These are strong recce elements that will fight, if necessary, to obtain information, but ideally they will try to surprise and defeat an enemy before they can respond. Combat recce patrols are drawn from the lead BMP battalion combined with some tanks and BRDMs.

Vanguard

The vanguard is formed by taking a company of BMPs from the lead battalion and reinforcing it with tanks from the tank battalion. The purpose of the vanguard is to provide reinforcement to the combat recce patrols if necessary, and to give early warning to the main body. It operates about 5 km in front of the lead battalion and includes a small detachment of BRDM-2 AT-5 anti-tank vehicles, which will operate as a group and provide a mobile anti-tank force.

Below: A later variant of the BMP is the BMP-2, mounting a 30-mm RARDEN-type cannon and the AT-5 'Spandrel' missile. It is not going to replace BMP-1, but is used in specialist platoons.

Tanks

The main Soviet tank in use in the Red Army of the 1980s was the T-64. The T-72, although a later model, was designed mainly for export and is less capable. Both are slowly being replaced by the new T-80, with its gas turbine engines and sophisticated armour.

Above: This is a Soviet-made T-72 of the Czechoslovakian army. The T-72 has a crew of four and no autoloader.

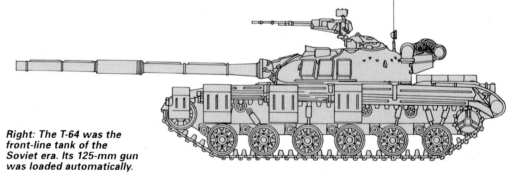

Right: The T-64 was the front-line tank of the Soviet era. Its 125-mm gun was loaded automatically.

Left: T-72 is used by the Russians, as can be seen by the up-armoured T-72 used in Afghanistan. The additional armour was a response to improved ballistic performance of NATO rounds and anti-tank missiles. Yet its armour is still far inferior to the advanced Chobham armour of a modern Challenger tank.

IN FOCUS

Artillery

At regimental level the main artillery piece will be the 122-mm self-propelled 2S1. The motor rifle regiment will have a regiment of artillery (three batteries of six guns). These are at the direct call of the regimental commander. He may, in addition, call upon divisional artillery, which is considerably more powerful, having rocket artillery and a vast amount of high-calibre conventional weaponry.

The 2S1 fires a 21.7-kg HE shell to a range of 15.3 km.

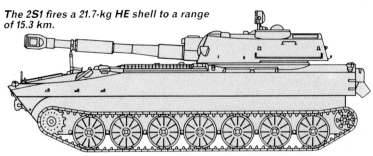

Above: At divisional level the 2S3 152-mm gun is the most common weapon. It can fire conventional, nuclear or chemical rounds.

The usual Russian tactic is to forward deploy a battery of six guns to provide direct support to the advance guard. The remaining guns are held back. However, if the regiment is leading a break-through, they will be reinforced with artillery.

 nk

ents
re

BMP battalions

Despite the attacking force of the tanks, they are fairly useless without the infantry. The soldiers in the BMP are the guys who will dismount, get in the enemy trenches and fight the battle. With the vast numbers of soldiers in a motor rifle regiment, it is a well-defended enemy that would be able to hold off this force.

135 km   145 km   145 km   150 km

Regimental rear security group

Since this is to protect an entire regiment, the group will be at least a company in strength. Their task is to provide cover and observation of the rear of the column.

Regimental rear march security patrol

Travelling at the back is one platoon of BMPs. Their task is to ensure link-up is made with regiments following and that the ground taken is handed over with no enemy re-infiltration.

The Russian Army is organised into two basic units: motor rifle regiments and ta regiments. The shows a typical formation that an advancing motor rifle regimen would adopt, but the exact formation would depend on threat and terrain. Regim do not operate on their own, but are combined to make divisions. These in turn combined to make armies and then fronts.

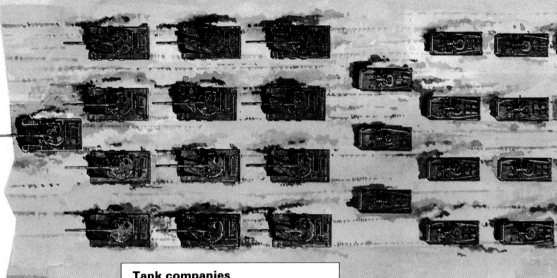

Tank companies
The bulk of the tank force will lead the BMP battalions in the main body. Made up of T-64, T-72 or even T-80 tanks, the companies provide the main shock action force of the motor rifle regiment. In an attack they will lead the BMPs on to the enemy and provide on-the-spot firepower. The tanks are the spearhead of the regiment.

70km   80km   100km

Regimental artillery group
Indirect fire support comes from the three batteries of 122-mm 2S1 guns. One battery is sent forward with the advance guard; the remainder are held with the regimental main body.

Regimental main body
Travelling in two columns about 30km long, 20km behind the advance guard is the bulk of the regiment. These are known as second echelon forces. Russian tactics put great emphasis on large forces being available to achieve a massive superiority in given area should they attack – on average, six or seven Russian tanks and APCs against every one of the enemy's. If the lead battalion

encounters an enemy force that it cannot deal with, the regimental commander will regroup his forces, use his lead battalion to hold the enemy in contact, and at the same time bring up this second echelon. Although this all takes time (about two hours to mount a regimental attack would be the very minimum), by completely swamping the enemy it actually saves time in the long run.

Right: The incredible ZSU-23/4 fires 23-mm ammunition at a rate of up to 1,000 rounds per barrel per minute with a maximum effective range of 2500 metres. If it fired at its full rate it would use up its ammo in half a minute! The engagement sequence can be fully automated. The radar will lock on at up to 15 km and report when the target is in range. There have been reports of the vehicle being used in the ground role against APCs and the like.

Left: ZSU-23/4s always operate in pairs for maximum protection. It is unlikely that they would be retasked from their primary role of protecting the commander, although they may possibly go forward with the advance guard if needed. One pair will always remain on primary tasking.

Inset below: The rather old and not terribly good SA-9 'Gaskin' could be used either with a separate radar tracker, or controlled optically. It was replaced by the vastly superior SA-13 'Gopher' system.

Air defence

The regimental commander's group will have its own air defence group protecting his vehicle. This will be provided by the ZSU-23/4 — a four-barrelled self-propelled gun that can fire at a rate of 1,000 rounds per barrel per minute. Protection for the entire regiment will come from its own SA-9 air defence guns, and it will also come under the umbrella of the divisional weapons.

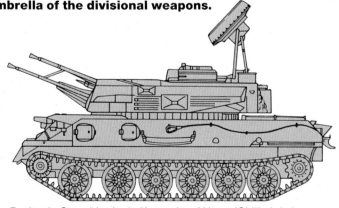

The Zenitnaia Samokhodnaia Ustanokva 23/4, or 'Shilka', is known to NATO as ZSU-23/4.

Mobile anti-tank

The best anti-tank weapon is another tank. However, in a motor rifle regiment the number of tanks is limited. In order to cover this deficiency a motor rifle regiment is supplemented by an anti-tank platoon. This is made up of nine BRDM-2 AT-5 anti-tank vehicles — the basic BRDM with the turret removed and five AT-5 'Spandrel' missile tubes mounted in its place. The weapon is accurate to 4000 metres.

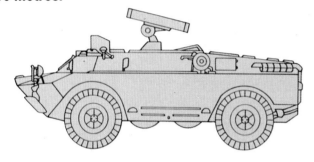

The AT-5 'Spandrel' is accurate to 4000 m. It takes 16 seconds to hit.

Recce

The BRDM-2 is the specialist recce vehicle of the regiment. Like the BMP, which is also used for recce, it is found in a large number of variants. There are special anti-tank and anti-air models, as well as a dedicated nuclear, chemical and biological recce vehicle. BRDM-2 has an extra pair of belly wheels that it can lower from inside if the terrain is particularly heavy going. This increases its cross-country speed.

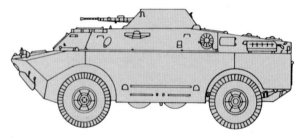

With only a 12.7-mm machine-gun, BRDM-2 will not fight for information.

Command vehicles

The command structure within the Soviet forces is very much more rigid than Western ones. Platoon commanders exercise no control over their platoon, and their vehicle radios cannot transmit, only receive. Company commanders and battalion commanders are not really expected to do much more than carry out orders; little initiative is expected of them. It is only the regimental commander, in his special BTR-60 PU Comd, who is actually required to do much in the way of tactical thought.

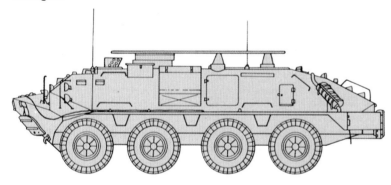

The controlling vehicle of the entire operation is the BTR-60 PU Comd.

Above: The Soviet army wisely invests a lot of effort in recce. Every organisation has dedicated recce groups, and it has entire recce regiments for large formations.

Below: The BRDM-2 is the eyes and ears of the motor rifle regiment. It will be tasked to check routes for the advance as well as establish enemy positions and strengths.

Main pic: Dismounted infantry fight alongside the Panzer III, the most numerous tank in Operation Barbarossa.

BLITZKRIEG! The Lightning War

Blitzkrieg was the first real example of modern mechanised warfare. It burst upon the world with the triumphs of the German army in the first two years of World War II, sweeping all before it; yet at the moment of greatest triumph it failed, in the snows outside Moscow in December 1941.

"At 0650 hours on June 22nd, 1941, I crossed the River Bug in an assault boat in the neighbourhood of Kolodno . . ."

Nothing odd about that, you might think – except that the writer was General-Oberst Heinz Guderian, commander of the Wehrmacht's Panzergruppe II, going into action in the opening stages of Operation Barbarossa as the German army swept into Russia. He was crossing the river ahead of most of the troops in his command, such was his commitment to the theory of Blitzkrieg and his own personal subscript to it: lead from the front.

"My command staff, consisting of two armoured wireless trucks,

a number of cross-country vehicles and some motorcyclists, followed at 08.30. I followed the tank tracks of the 18th Panzer Division and soon reached the bridge over the Lesna, whose capture was important for the advance; there I found nobody except for some Russian pickets . . . at 10.25 the leading tanks reached the Lesna.

"We had taken the enemy by surprise along the whole Panzer Group front . . . the enemy, however, soon recovered and put up a tough defence in his prepared positions. Brest-Litovsk held out with remarkable stubbornness for several days.

"In the evening, the Panzer Group was fighting around Malortya, Kobryn, Brest and

EYE WITNESS

"The flood burst into France. From each of the bridgeheads the Panzers roared out, preceded by a cloud of screaming Stukas. They were covered from our fighters by Messerschmitts. As we pushed south, refugees clogged the roads. The poor people were harried by German fighters, bullied by frightened and demoralised French soldiers and gendarmes, or forced into ditches by strange, ominous, foreign vehicles manned by blond giants who waved triumphantly at them. Of course, they were not all blond, nor were they giants, but as they forced their way down the roads, leaving an impression of total invincibility in their wake, it just seemed that way. Churchill flew to Paris the next day and asked the French where their reserves were, only to be told that there were none, and that they had been defeated."

Lieutenant, British Expeditionary Force, France 1940

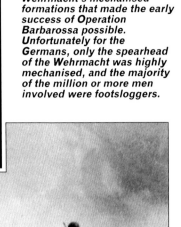

Below: A German Panzer III is accompanied by Panzergrenadiers as it moves across the vast expanses of the Russian plains. It was the speed of advance of the Wehrmacht's mechanised formations that made the early success of Operation Barbarossa possible. Unfortunately for the Germans, only the spearhead of the Wehrmacht was highly mechanised, and the majority of the million or more men involved were footsloggers.

Colonel-General Heinz Guderian, the architect of Blitzkrieg, passes instructions to the communications technicians in his half-track during an advance.

Pruzana, and it was at this last-named that the 18th Panzer Division became involved in the first tank battle of the campaign."

On that first day, Guderian's Panzers had advanced more than 50 kilometres in a corridor that followed the road east from Brest. In the course of the next week, they were to keep up that daily advance, and also spread out until they formed a two-pronged salient 400 kilometres deep through Byelorussia and into the Ukraine.

Weak spot

This was mechanised warfare at its purest and most successful: finding a weak point, punching through it aggressively, leaving strong pockets of the enemy to be 'mopped up' by the second echelon troops coming up behind. Such was the scale of the invasion of the USSR that those 'pockets' were often large cities defended by tens of thousands of troops!

The Blitzkrieg theory evolved out of the German infiltration tactics used at the end of World War I, when specially trained troops would slip through the enemy defences to wreak havoc in the enemy rear. Success depended on speed, but troops on foot, no matter how fit or well-trained, are going to reach exhaustion point sooner or later.

Blitzkrieg, developed by Guderian and others in the inter-war years, was the infiltration technique used on a large scale. It depended upon mechanisation, with armour, infantry, and artillery all moving at high speed into the enemy rear area.

It had seen its first tests two years before, against much softer opposition. In 1939, Guderian's Panzers had been the spearhead of Hitler's invasion of Poland, and in 1940 it had been the mobile forces striking through the Ardennes which had dislocated the Allied armies moving north into Belgium, forcing the British off the Continent and the French into eventual capitulation. Russia was to be its real test, however, as the distances and extremes of climate were to demand the most from men and machines.

Advancing in Russia

Amongst the men doing the actual fighting, the initial success of the invasion of the Soviet Union was intoxicating. This account, by Helmut Pabst, a 30-year-old NCO, tells of the advance on Smolensk, at the end of June:

"Number 10 Battery was dug in in a potato field. We were to open fire at 03.05 hours. At the same moment, everything sprang to life. There was firing along the whole front. The Russian watchtowers vanished in a flash and shells crashed down on the enemy batteries, located long before. In file and in line, the infantry swarmed forwards, through bog and ditches, boots full of mud and water. Ahead, the barrage crept forward from line to line. Flamethrowers advanced against the strongpoints, their roar competing with the fire of machine-guns and the high-pitched whip of rifle bullets. At Kanopky Barracks came the first serious resistance, and the company ahead stuck. Assault guns forward! came the order.

"The advance went on. We moved fast, sometimes flat on the ground, but irresistibly. Ditches, water, sand, sun. Always changing position; thirsty and hungry, with no time to eat. By 10 o'clock we were already old soldiers and had seen it all: abandoned positions, knocked-out armoured cars, the first prisoners. The first dead Russians.

"In the ruins of the first burnt-down village, where only the chimneys still stood, we came under shellfire for the first time. The shells make a curious singing noise; you dig in fast, and make yourself flat.

"About three o'clock the next morning we got through the dugout line. Suddenly, we came to a halt. Someone called 'Anti-tank guns - forward!' The guns raced through. The next two kilometres was a waste of sand, covered with clumps of broom and, beyond, the fortress of Osowiec, with our forward positions on the riverbank. The Russians knew exactly where we were . . ."

Pabst's father had fought in World War I, and in his letters to him his son tried to highlight the differences between that and 'his' war:

A different war

"Of course, we're up against different weapons, but we have different ones ourselves. A tank can be awkward if you only have your rifle, but you can always squeeze into a hollow and let it pass. And even a brute like that isn't invulnerable to a single man, provided you can get it from behind.

"Artillery and infantry still dominate the battlefield. The increased firepower of the infantry

– automatic weapons, mortars and the rest – isn't as bad as it's made out to be. But you have to accept the fact that you're after the other man's life. That's war. That's the trade. And it isn't so difficult. Again, because the weapons are automatic, most people don't realise the implication: you kill from a distance, and kill people you don't know and don't see. Situations in which soldier confronts soldier are possibly more common in this campaign than in the previous one [World War I], but they don't happen that often."

The pitch of activity Pabst describes seems impossible to maintain, but the two million men of the German army did just that throughout the summer months, snatching their rest where they could, and measuring it in minutes rather than hours.

By 20 July, a month after the attack was launched, the main thrust had covered 800 kilometres (500 miles), but it wasn't enough.

The logistics of such a rapid advance were nightmarish. The initial German attack was made with 18 Panzer divisions, 12 motorised infantry divisions and no less than 80 infantry divisions, with a further 25 divisions in

reserve. Two million men, 3,200 tanks and 10,000 guns were supported by massive stores, fuel and ammunition dumps on the Soviet border, enough for a 650-kilometre advance. 500,000 trucks and 300,000 horses had been gathered to move the

Below: German experiments in battlefield mobility dated back to well before the war. Here, members of the NSKK (Nationalsozialistisches Kraftfahrer Korps, the motor-transport corps of the Nazi party, later absorbed into the SS) use the hulls of Panzer I light tanks as personnel carriers.

Left: The first real test of Blitzkrieg tactics came in France in 1940, as the invasion of Poland was a conventional operation. German tanks burst through the Ardennes, followed by fast-moving, motorised infantry, and the sheer speed of their advance disrupted French defences.

Right: German troops advance into Russia in June 1941. The invasion was supported by huge supply dumps in occupied Poland, and half a million trucks were used to ferry those supplies forwards.

Above: Mechanised troops were only the leading part of the massive German invasion. As they had earlier in the war, infantry and horses played a major part in the German plans.

Below: One thing the Germans did not count on was the Soviet winter, and another was the strength of the Soviet counter-attack before Moscow. Blitzkrieg had failed.

supplies forwards.

But the mechanised formations had reached that far into the USSR in three weeks, and a supply organisation that still relied heavily on horse-drawn transport would be hard pushed to maintain the Panzers at such a rate of advance. Mechanising your combat arms allows you to advance at a great rate, but if your supplies cannot catch you, then the movement which is such an essential part of Blitzkrieg must grind to a halt.

As the supply lines got longer, the advance bogged down, and that was precisely what the German high command could not afford. They had to capture Moscow before winter froze them in their tracks, and like Napoleon before them they got to within sight of the Kremlin spires.

But it wasn't just 'General Winter' that held the Germans. The Soviets had traded territory for time, building new defences well back from the enemy, and wearing the invaders down on their long advance. The casualties in the first year of war on the Eastern Front were appalling: nearly a million Germans killed, wounded or missing, and many times that number of Soviet casualties or taken prisoner. But with the failure of the motorised drive on Moscow, the war settled down into one of attrition. Never again would Blitzkrieg come so near to success: never again would mechanised warfare ultimately fail so disastrously.

RIDING INTO BATTLE

The armoured personnel carrier is bullet-proof and protects you from shell splinters, but an anti-tank missile will smash through, with horrific consequences.

Built from special aluminium alloy that keeps its all-up weight to less than 11 tonnes, the M113 APC gives its occupants limited protection against small-arms fire and from grenade and shell fragments, but not from anti-tank guns, missiles and rockets. This means that the infantry leader must think very hard about when and how to dismount his troops and use them in the traditional foot soldier's role.

In the attack, the leader will try to fight as long as possible, using tactics thought out in advance for just this sort of situation. He will get his men out of the APCs only when in close terrain such as wooded areas, or if he comes up against a strong anti-tank force.

This flexibility – to fight from vehicles with armour to defeat small arms, or dismount and take on anti-armour forces with conventional infantry tactics, then perhaps call up your APCs, mount up again and carry on the advance as before – has meant that a new way of fighting has had to be developed.

It is unlikely that mechanised infantry will operate in an area without tank support, and vice versa. They are dependent on each other; areas that are unsuitable for tank warfare are equally unsuitable for mechanised infantry. However, it is possible that the APCs may be used to take a force to the edge of an area so that the infantry can dismount and fight in the conventional infantry role.

1 Travelling

The method of travel will be dictated by the three Ts: Terrain, Threat and Time. If all three are low – easy terrain, low threat and plenty of time – then the infantry platoon will move as a single block, either in a line or perhaps staggered line ahead. If the threat is slightly higher, travelling overwatch is likely to be used. A scout group is sent out ahead within range of weapons to clear the route. Both groups are moving together, but the scouts are kept under the protective envelope of the main body behind. If the threat is high, bounding overwatch will be needed. The group is split into two. One group remains firm while the other moves. Once the first group is firm the second group will move, covered by the first, and so on.

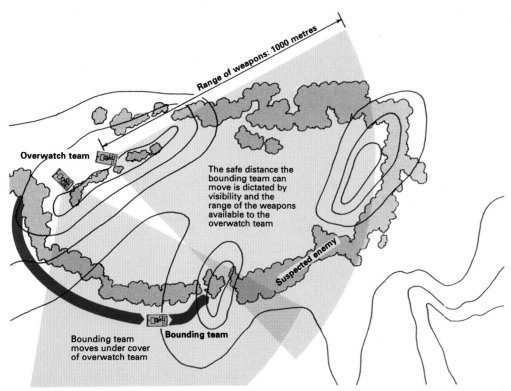

Range of weapons: 1000 metres

Overwatch team

The safe distance the bounding team can move is dictated by visibility and the range of the weapons available to the overwatch team

Suspected enemy

Bounding team moves under cover of overwatch team

Bounding team

Above: Introduced in 1964, the M113 is only now
being replaced by the Bradley M2. It is unlikely
that the M2 will ever achieve the sort of exports
and incredible number of variants that the M113
generated.

2 Advancing along roads

The problem of using roads in areas of high threat is
that they are very easy to ambush and mine. Before a
force can travel down them, they must be cleared; a
slow and laborious procedure. The APCs will get in a
position from which they can cover the dismounted
force sent forward to explore. The dismounted teams
will first clear the high ground, looking for enemy
positions or control wires. At the same time another
party will be clearing the road, checking for booby
traps, mines etc. Once they report clear, the APCs can
move forward to a new position to cover the force as
it moves on to clear the next stretch.

Above: The second-
largest user of the
M113 is the Israeli
army, which knows it
as 'Zelda'. Recent
modifications
include the addition
of spaced armour to
defeat anti-tank
missiles.

Below: The M113 was
widely used in
Vietnam. The squad
often rode on the top,
which lost the small-
arms protection but
meant that if an RPG
hit the casualties
would be reduced.

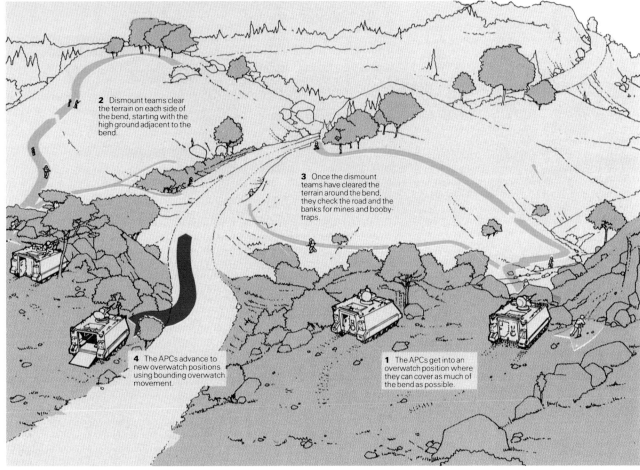

2 Dismount teams clear the terrain on each side of the bend, starting with the high ground adjacent to the bend.

3 Once the dismount teams have cleared the terrain around the bend, they check the road and the banks for mines and booby-traps.

4 The APCs advance to new overwatch positions using bounding overwatch movement.

1 The APCs get into an overwatch position where they can cover as much of the bend as possible.

TACTICS

Securing a bridge or defile

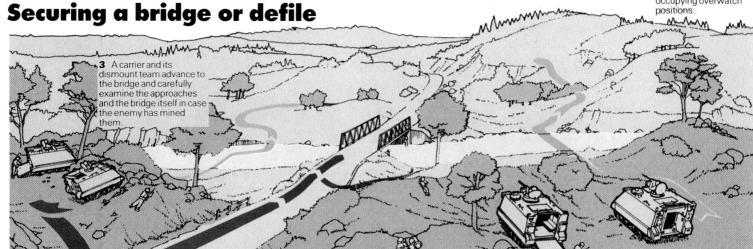

3 A carrier and its dismount team advance to the bridge and carefully examine the approaches and the bridge itself in case the enemy has mined them.

2 If a fording site is available the dismount teams cross the river and secure the far side, occupying overwatch positions.

1 The APCs move into an overwatch position where they can cover the terrain on both flanks of the bridge and the far side.

3 Bridge or obstacle crossing

Bridges or defiles (driving across a gap such as a valley bottom) must always be treated with great caution. Being an obvious bottleneck, they are very likely to be ambushed. Again the principle of cover applies. By positioning the APCs so that they can cover the bridge or obstacle gap, they will be able to bring down quick and accurate fire. Perhaps the best way of dealing with a bridge is to make use of the M113's swimming ability and find a crossing site. Once you have got troops over to the other side it greatly increases the security of the operation.

Above: The temptation, when you spot a bridge like this, is to go hell for leather at it. Fine if you know it's clear. But how certain are you?

Below: To an anti-tank missile controller, an APC full of men is an ideal target. You must know exactly what to do if you are targetted.

4 Dodging anti-tank missiles

Anti-tank missiles are a terrible threat, and a hit on a fully loaded APC would be catastrophic. In order to hit the APC, the enemy missile controller must keep the crosshairs of his sighting system on your APC. One way of reducing the threat is to drive in an erratic way, speeding up, slowing down and taking sharp turns. This may be unpopular with those in the back, but it's preferable to being hit by a warhead. If you spot a missile launch, your immediate reaction should be to return with massive amounts of machine-gun fire in the direction of the launch. It may not hit him, but it may cause him to break his aim.

Missile right!

1 Look out for the tell-tale flash as the enemy anti-tank guided missile operator launches his missile.

2 Immediately fire all your machine guns at the enemy gunner: if it does not hit him it will at least disturb his aim.

5 The assault

The commander must make the decision on when to dismount his infantry and fight through the position. If the enemy is a weak one and has not had time to prepare, he may be confident enough not to dismount them at all, in which case the APCs will steamroller through the position, putting down fire from the vehicle's weapons. However, the risk from hand-held anti-tank weapons is very high. It is much more likely that the commander will dismount his infantry on the forward edge of the enemy position. This will enable them to storm the position, being supported by the vehicle weapons, but the vehicles will not be threatened by the anti-tank guns.

Dismounting in contact is very confusing. You have no idea where the enemy is or what is going on.

SELF-PROPELLED ANTI-AIRCRAFT GUNS

The modern conflict is now the air-land battle. For the forces stuck to the earth's surface, the ability to inhibit or destroy the aerial attackers is a vital function. Self-propelled anti-aircraft guns are essential if the ground troops are to survive.

It was May 1940. The German armoured spearheads had swept across the borders of the Low Countries. The Allies pushed into Belgium in a pre-arranged manoeuvre to head off the German attack. The move was supported with the AASF (Advanced Air Striking Force), made up of the Bristol Blenheim and Fairey Battle bombers of the Royal Air Force. Slow, and easy meat for the fighters of the Luftwaffe, the bombers flew their missions at low level. In spite of suffering heavy losses, the pilots of the AASF were keen to hit at the invaders.

German capture of the two bridges over the Albert Canal at Maastricht threatened the whole Allied position. The High Command decided to cut the bridges. Belgian artillery and French high-level bombing had failed to have any effect. Six Fairey Battles manned by volunteers from No. 12 Squadron were ordered to attack at low level. One failed to take off, leaving five for the mission.

As each Battle came within range, it was met by a curtain of tracer. The Germans had moved numerous light flak batteries in to defend the bridges as soon as they were taken. Riddled with cannon shells, every lum-

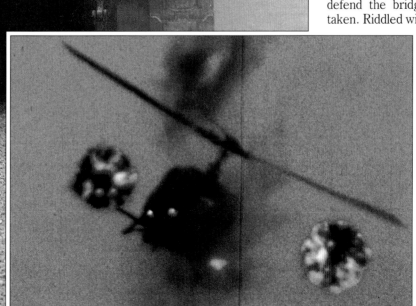

The Italian OTO-Melara quad 25-mm (SIDAM 25) system. One of the most recent developments in the SPAAG field, it shows a typical technique – multiple, fast-firing guns on a single mount – in this case, 25-mm cannon firing at 2,400 rounds per minute. Inset: The receiving end of anti-aircraft fire. In this case, a UH-1 helicopter is brought down by proximity-fused rounds. Small-calibre weapons usually have to actually hit the target to score a kill. Larger-calibre rounds can use more sophisticated fusing techniques.

bering bomber was shot down. But this was not the end of the story. When the main German offensive burst through the Ardennes a few days later, the Battles were sent to attack the enemy pontoon bridges across the Meuse. In the course of one hour, 28 out of 37 attacking aircraft were destroyed. Never before or since has the Royal Air Force sustained a higher proportion of casualties to such little effect. The highly trained and highly mobile German flak crews had ensured that the Wehrmacht's Panzers could continue their advance unchecked.

To many people, mechanised warfare means tanks and armoured infantry vehicles.

This ignores the other essential components of a mechanised force, such as engineers, artillery, and particularly air defences. These have become more important as the rise in tactical air power has added a new dimension to the modern battlefield. Aircraft can deliver sophisticated, accurate weaponry in high-speed, low-level attacks, and defences have had to evolve to meet the threat. Sometimes the aircraft have the advantage, while at others, most notably in Belgium in 1940, it is the guns which very definitely come out on top.

Self-propelled anti-aircraft guns (SPAAGs) are by no means new. Before the start of

World War I, armoured trucks with guns had been developed to engage the balloons used for artillery spotting. It was not until after the war, however, that serious work on SPAAGs began; it took World War II to show that they have a place in battle. Large numbers were built by the major powers, and were fielded on wheeled, half-track and full-track chassis.

Gun development

Many of the early guns had a calibre of 20 mm, studies in the 1920s having shown that this was the largest shell capable of carrying a reasonable explosive charge that would be economic to fire automatically. As World

In an attempt to build a coherent air defence umbrella following the collapse of the Sergeant York programme, the US have trialled and introduced the Swiss ADATS missile system. Using a tracking radar and a coded CO_2 guidance laser, the rocket travels on an unjammable beam at Mach 3. The US variant (shown here) has a 25-mm Chain Gun added to the missile mount for use at close range, or when all missiles have been fired.

With its twin 35-mm cannon and Marconi 400 radar, the Marksman was an interesting project. The system was designed to be entirely self-contained and the turret could be mounted on almost any chassis; it was trialled on the T-55, M84, Chieftain, Centurion, Challenger, and Vickers Mk 3.

World War I AA guns

FLASHBACK

Although aircraft were in their infancy, towards the end of the Great War they did pose a significant threat to ground forces. The history of the anti-aircraft gun is the history of the aircraft. As the aerial threat developed, it became necessary to counter it. Initially this was little more than small-arms fire from the troops, but it became apparent that more was needed. Various experiments were tried, using high-elevation artillery pieces. Calibres ranged from 20 mm up to 76 mm. They were perhaps of more use as a deterrent than for actually scoring hits. The prospect of a large amount of flak was sufficient to deter all but the most cavalier pilots.

American troops on the Western Front man an anti-aircraft version of the famous French-built 75-mm cannon.

SELF-PROPELLED ANTI-AIRCRAFT GUNS Reference File

227

FORMER USSR

ZSU-30/4

Formidable though it still is, the ZSU-23/4 is being supplanted by the yet more capable **ZSU-30/4** system, with the Russian service name and production designation **Tunguska** and **2S6** respectively. The ZSU-30/4 is a hybrid type combining four extremely powerful 30-mm barrels and no fewer than eight SA-19 surface-to-air missiles. The system entered service in the late 1980s, and is designed for protection of high-value battlefield formations against aircraft flying at up to 1800 km/h (1,118 mph) at heights between 15 and 3500 m (50 and 11,485 ft).

The welded-steel turret for the weapon system is located on a welded-steel chassis based on that of the MT-S

tracked command vehicle. The weapons are located on the outer sides of the turret, so avoiding the problem of gun gas in the fighting compartment, and comprise four 30-mm barrels (generally equivalent to Western 35-mm weapons), and either two closely located pairs or two twin-barrel weapons) and, outside these, two groups of four SA-19 SAMs; the missiles have an elevation system separate from that of the cannon. Above the turret rear is the 'Hot Shot' radar, whose separate acquisition and tracking elements allow continued surveillance while any target is being tackled. There is also a back-up optronic fire control system.

Specification
ZSU-30/4
Type: four-crew self-propelled 30-mm AA gun/SAM system
Weight: 34 tonnes (33.46 tons)
Armament: four 30-mm cannon with between 250 and 350 rounds per gun, and eight SA-19 SAMs

Performance: maximum speed 65 km/h (40.5 mph); range 500 km (311 miles)
Dimensions: length overall 7.84 m (25 ft 9 in); width 3.47 m (11 ft 5 in)
Powerplant: one 390-kW (523-hp) V-59 (modified) diesel engine
Users: Former USSR and Warsaw Pact

Left: The mainstay of the US air defence umbrella is still the Vulcan. It was originally introduced in 1964 but has recently undergone improvements to both the fire control system and radar, which has improved efficiency at the same time as simplifying operation – something of a first for the Americans. The original aircraft version fired at 6,000 rounds per minute, but that has been reduced to a maximum of 3,000 rounds per minute.

War II progressed, however, firepower had to be increased to deal with faster, better-protected aircraft. One way to boost the weight of fire was to field bigger weapons, like the highly successful Bofors 40-mm. Smaller-calibre weapons were also improved by increasing the number of barrels that could be fired simultaneously from a single mount.

The first post-war systems were not much of an advance over their predecessors. The American M42 'Duster' mounted twin 40-mm cannon on a light tank chassis, but its optical sights limited it to clear-air, daylight operations, and its manual controls made it difficult to engage fast-flying targets. The M42 was used extensively in Korea and Vietnam, but mainly in the ground support role.

As aircraft got faster, it became clear that the M42 could not cope. The awesome Vulcan

The Professional's View:

Arab air defences

"**W**e were ready for the SAMs, having been in against them before. The older SA-2s were pretty easy to deal with and our ECM pods were ready for them. When we got there, though, it was altogether different. Our int[elligence] boys let us down. Sure enough, the SA-7s were coming up at us as we had pretty much expected, but it was the SA-6 that took us by surprise. The ECM pods had no effect and they were big and damn mean. You could outmanoeuvre them if you were lucky, but it was rough.

"We tried to drop low to avoid them. That was when we ran into the ZSUs. Before, they hadn't been too much of a problem. Fly high enough and their range is too low to get you. Now we were trapped. If you flew too high, the SA-6 got you. If you dropped too low to avoid them, you ran into the ZSUs. Anything that can pump out that many shells in that short a time is just best avoided."

Israeli F-4 Phantom pilot

228

USA

FMC/General Dynamics M163 Vulcan

The **M163A1 Vulcan** AA gun system was accepted for US service in 1968 and is little more than the M167 towed AA gun mounted on the modified chassis of the M113 armoured personnel carrier. The type therefore possesses useful battlefield mobility, but retains the tactical limitations of the M167. These limitations are a small-calibre projectile, restriction to clear-weather operation, and an open mounting that precludes any form of NBC protection.

The weapon is the 20-mm General Electric M168, a rotary six-barrel cannon modelled on the US Air Force's M61A1 Vulcan aircraft weapon. The M168 has an effective range of 2970 m (3,250

yards) against surface targets firing at 1,000 rounds per minute, and of 1600 m (1,750 yards) against aircraft firing at 3,000 rounds per minute; the ammunition types are HE incendiary and armour-piercing. The mounting is electrically powered through 360° traverse, and the cannon can be elevated in an arc between −5° and +80°. Even if warned by a radar system, the gunner has to acquire his target visually and track it with the Mk 20 lead-computing sight, range data being supplied by VPS-2 radar.

Specification
FMC/General Dynamics M163A1
Type: four-crew self-propelled 20-mm

AA gun system
Weight: 12.311 tonnes (12.116 tons)
Armament: one 20-mm multi-barrel cannon with 2,100 rounds
Performance: maximum speed 68 km/h (42 mph); range 483 km (300 miles)
Dimensions: length overall 4.86 m

(15 ft 11 in); width 2.85 m (9 ft 4 in)
Powerplant: one 160-kW (215-hp) Detroit Diesel 6V-53 diesel engine
Users: Ecuador, Israel, Japan, Jordan, Morocco, Portugal, Saudi Arabia, South Korea, Sudan, Thailand, Tunisia, Yemen and USA

The expectation of the 1980s that high-energy lasers would replace existing SPAAG vehicles has not been realised.

20-mm cannon, developed to arm supersonic fighters, was adapted for the anti-aircraft task. Fitted with a naval gunsight and a ranging radar, it was a considerable improvement on the M42 in some respects, although in others it was a distinct step backwards, having shorter range and firing considerably less lethal rounds.

In an attempt to improve the situation, the US Army developed DIVAD, the Divisional Air Defence system. This used twin 40-mm cannon mated to a radar adapted from the F-16 fighter, mounted on an M48 tank chassis. Unfortunately, the project proved to be a classic example of how not to produce a weapon system, and after repeated and costly failures it was cancelled.

US and Soviet air defence

In a major re-assessment of air defence requirements, the US Army has evolved the Forward Area Air Defense System. This is an integrated programme of complementary gun and missile systems. The LOS-R (Line Of Sight – Rear) system is a combination of 0.5-calibre machine-guns and Stinger missiles mounted on a Hummer wheeled vehicle. The LOS-F (Line Of Sight – Forward) system utilises the Swiss-developed ADATS missile, with a 25-mm Chain Gun co-axially mounted.

The Soviets followed a similar route, first fielding a clear-weather system in the shape of the ZSU-57/2, followed by the all-weather ZSU-23/4. The twin 57-mm guns of the earlier weapon were considerably harder-hitting than Western equivalents, and the power-

---

Cone of fire

The guns can elevate from -4° to +85°

+85°

The weapon has a blind area of 10° immediately above it

Range of fire is 3000 metres (3,281 yards)

0°

-4°

The turret can traverse through 360°

ZSU-23/4 anti-aircraft gun

-4°

The ZSU-23/4 is perhaps the quintessential self-propelled anti-aircraft gun. Its quad mounting, high-firing gun system with radar tracking, and servo-electric turret are to be found in numerous manifestations throughout the world on any number of systems. The ZSU-23/4's cone of fire is impressive. With the exception of a 10-degree 'dead spot' directly above the weapon, it can engage in a complete hemisphere out to between 2500 and 3000 metres. It has the ability to operate as a stand-alone system or can be linked to other air defence radar systems. Using its own radar, it can acquire and identify targets out to 20 km.

---

229 — OTO-Melara Otomatic

ITALY

The **Otomatic** is unique among modern self-propelled AA gun systems in having just one large-calibre weapon, a 76-mm (3-in) ordnance derived from the OTO-Melara Super Rapid faster-firing version of the widely-used Compact naval equipment. The gun is located in a welded-steel turret located on the hull of the same company's OF-40 battle tank, and is used with turret-mounted acquisition and tracking radars. The manufacturer's rationale for the use of this weapon is simple but compelling: the system offers the capability of a medium-range SAM system without the considerably greater cost of the missiles.

The gun's rate of fire (120 rounds per minute) and accuracy allow a burst of five or six proximity-fused projectiles to be delivered right into the target's flight path at a range greater than that of the target's anti-tank missiles. This offers the probability of destroying the target before it can engage the armour being protected by the Otomatic, and without the cost of the two or more SAMs that would otherwise have to be used. The turret is powered for 360° traverse, the gun can be elevated in an arc between −5° and +60°, the limited elevation angle being sufficient for engagements out to 6000 m (6,560 yards), and an automatic feeding/loading system is fitted.

Specification
OTO-Melara Otomatic
Type: four-crew self-propelled 76-mm (3-in) AA gun system
Weight: 46.836 tonnes (46.1 tons)
Armament: one 76-mm (3-in) gun with 90 rounds, one 7.62-mm (0.3-in) machine-gun with 4,000 rounds

Performance: maximum speed 60 km/h (37 mph); range 600 km (373 miles)
Dimensions: length overall 9.376 m (30 ft 9 in); width 3.35 m (11 ft)
Powerplant: one 746-kW (1,000-hp) Fiat V-12 MTCA diesel engine
User: none as yet

---

230 — Marconi Marksman

UK

Unlike the other systems in this batch, the **Marksman** is not a complete vehicle but rather a turreted weapon system designed by Marconi Command and Control Systems for installation on any main battle tank with a turret ring of the appropriate diameter. This offers the operational, maintenance and economic advantages of re-using tanks which have become obsolete in their armament rather than through any failing of the hull and powerplant combination. The Marksman turret has been ordered by Finland for use on Soviet T-54/55 chassis.

The turret is of welded-steel construction, and the armament is the same combination of two 35-mm KDA cannon used in the Gepard's Oerlikon-Bührle GDP-C02 turret. The cannon are externally mounted, and can be elevated through an arc between −10° and +85°; the ammunition for each cannon is held in a container, and this simplifies resupply. The fire control system uses a Marconi 400-series lightweight acquisition and tracking radar that can find targets out to a range of 12000 m (13,125 yards), and an automatic computer fire control system requires the gunner only to press the 'fire' button. The turret also has stabilised optical sights.

Specification
Marconi Marksman
Type: turreted 35-mm AA gun system
Weight: 11 tonnes (10.83 tons)
Armament: two 35-mm cannon with 500 rounds, and four smoke-dischargers on each side of the turret

Performance: not relevant
Dimensions: not revealed
Powerplant: not relevant
User: Finland

WEAPON FILE

Left: As another private venture, but with an eye to a very lucrative US Army contract, General Electric produced a number of turrets to fit the M3 Bradley, including this combination with four Stinger missiles and a 25-mm GAU-12/U cannon. The radar will be dual role: acquisition and detection for both systems, and guidance for the cannon. Stinger missiles are heat seeking and require no guidance.

Below: The 76-mm shell from the Italian Otomatic gun has a burst radius sufficient to cause damage over a wide area, thus reducing the number of rounds that need to be fired. The greater range of the Otomatic, compared with small-calibre systems, also enables it to deal with helicopters making stand-off missile attacks.

operated traverse and elevation made the crew's job less difficult. The four-barrelled ZSU-23/4 has seen extensive combat since its introduction in the 1960s, and has proved one of the most effective battlefield weapons ever built, giving the Israelis quite a shock when used in conjunction with SAMs during the 1973 Yom Kippur War. Its radar is effective out to 20 kilometres in search mode, and can track targets and control the four 23-mm guns

231
FRANCE

Panhard M3 VDA

The Panhard AML (4×4) light armoured car has been one of the most successful of its type designed since World War II. To operate with this vehicle, Panhard designed the 4×4 APC which uses 95 per cent of the components of the original vehicle. This vehicle, called the M3, was equally as successful and, since production started in the early 1970s, over 1,500 have been built for export to over 30 countries. The anti-aircraft member of the variants is the **M3 VDA** (Véhicule de Défense Antiaérienne). It entered production in 1975.

The VDA is a standard M3 APC on which a turret with twin 20-mm Oerlikon cannon has been mounted. It has a crew of three with the driver at

the front, the gunner in the turret (in the centre of the hull) and the commander at the rear. The turret has full powered traverse through 360° at a speed of 60° per second. Mounted on the turret rear is a tracking and surveillance radar scanner that rotates at 40 rpm.

The gunner can select single shots, bursts or full automatic fire. Two cyclic rates are available: 200 or 1,000 rounds per minute. Each cannon is provided with 300 rounds of ready-use ammo, and additional ammo can be carried in the hull.

Specification
Panhard M3 VDA
Type: three-crew self-propelled twin

20-mm AA gun system
Weight: 7.2 tonnes (7.09 tons)
Armament: two 20-mm cannon with 300 rounds per gun; one 7.62-mm (0.3-in) machine-gun with 200 rounds and two smoke grenade-dischargers
Performance: maximum speed 90 km/h (56 mph); range 1000 km (621

miles)
Dimensions: length overall 4.45 m (17 ft 5 in); width 2.4 m (7 ft 11 in); height 2.995 m (9 ft 11 in)
Powerplant: one 67-kW (90-hp) Panhard model 4 HD air-cooled petrol engine
Users: Ivory Coast, Niger and the UAE

232
GERMANY/FRANCE

Thyssen-Henschel/Thomson-CSF Dragon

Given the ever-rising cost of defence equipment, there has been an increasing trend for co-operative weapons systems development. Private venture development has also gone along similar lines, and the product of just such a venture was the **Dragon** twin 30-mm SPAAG. The chassis was developed by Thyssen-Henschel, then of West Germany, while the turret and associated fire-control systems were developed by the electronics division of Thomson-CSF of France.

The all-welded steel turret is mounted in the centre of the hull, with the commander seated on the left and the gunner on the right. Mounted on the rear is the 'Oeil Vert' ('green eye')

radar that carries out both tracking and surveillance functions. When not required, the radar can be retracted into the turret bustle.

The turret has full powered traverse through 360° at a speed of 35° per second and the twin 30-mm cannon can be elevated from −8° to +85° at a speed of 30° per second. All controls are hydraulic with manual emergency back-up.

The gunner can select bursts from one to five rounds or a burst of 15 rounds. The weapons have a maximum effective range of 3000 metres and can also be used against ground targets.

Developed specifically for export, the Dragon found no buyers.

Specification
Thyssen-Henschel/Thomson-CSF Dragon
Type: three-crew twin 30-mm AA gun
Weight: 31 tonnes (30.5 tons)
Armament: two 30-mm cannon with 2,100 rounds
Performance: maximum speed

72 km/h (45 mph); range 600 km (373 miles)
Dimensions: length 6.775 m (22 ft 3 in); width 3.12 m (10 ft 3 in); height 4.195 m (13 ft 11 in)
Powerplant: one 536-kW (720-hp) MTU-6 supercharged diesel engine
User: No buyers

at ranges of up to eight kilometres.

The most recent Russian SPAAG is the 2S6, also known as the ZSU-30/2, which combines four 30-mm cannon with a number of short-range, heat-seeking missiles. Russian regiments are now equipped with six SA-13 'Gopher' SAM vehicles in place of four SA-9s, and six ZSU-30/2 vehicles in place of four ZSU-23/4s. The increase in numbers, together with the greater range of both guns and missiles, mean that the air defence battalion of a regiment can now defend twice as much territory as before.

Recent weapons development

Experience of being on the receiving end of Allied air power during World War II taught the Germans the value of good air defences. The Gepard is a potent air defence system, developed in association with Contraves of Switzerland, which has twin 35-mm cannon and a comprehensive electronics suite

mounted on a Leopard tank chassis.

Small-calibre cannon are all very well, but one of the major threats to armour on the modern battlefield is from helicopters with long-range missiles. They are able to mount stand-off attacks from outside the range of 35-mm or 40-mm cannon. The Italian solution is to use a quick-firing naval 76-mm gun on a tracked chassis. The Otomatic has an effective anti-aircraft range of six kilometres, and

its 12-kg round is more than 20 times heavier than a typical 30-mm shell. The Italians have also developed the close-range SIDAM 25. This is a quadruple 25-mm cannon similar in concept to the Soviet ZSU-23/4.

France has developed a number of systems in recent years, although the French army makes only limited use of SPAAGs. A large proportion of French-manufactured equipment is for export. Swiss companies are

In its lifetime the ZSU-23/4 has undergone numerous upgrades and refits. The first models had analogue computers and a rather crude radar system; both features have since been updated. As a result the ZSU, despite its advanced years, is still an effective and battle-proven system. To its detriment, however, is its thin armour and lack of any amphibious capability.

Combat Comparison

233 FORMER USSR

ZSU-23/4 Shilka

Representing the best from France and Russia, the ZSU-23/4 and the Gepard are unlikely now to find themselves pitted against each other. However, they typify two approaches to the same problem.

In numerical terms, the **ZSU-23/4 Shilka** is still one of the most important self-propelled AA gun systems in the Russian inventory. As suggested by is designation, it has an armament of four 23-mm cannon. Soviet cannon are generally superior to Western ones of the same calibre, firing a heavier projectile with a higher muzzle velocity, so the ZSU-23/4 can be reckoned to be the equivalent of a Western system with quadruple 30-mm cannon.

The equipment entered service in the mid-1960s, and is based on the chassis of the PT-76 amphibious light tank. The large but low turret is also made of welded steel, and accommodates at its rear the 'gun dish' radar that can acquire targets at a range of 20 km (12.4 miles) and track them from a range of 8 km (4.97 miles). Each cannon is water-cooled and has a cyclic rate of between 800 and 1,000 rounds per minute with HE and armour-piercing ammunition, each with an

incendiary component. The cannon have an effective range of 2500 m (2,735 yards), and the high rates of traverse and elevation (the turret through 360° and the cannon in an arc between −4° and +85°) allow the effective engagement of fast-flying and fast-crossing targets even at short range. The type has been produced in steadily improved variants, and is still a formidable weapon.

Specification
ZSU-23/4
Type: four-crew self-propelled 23-mm AA gun system
Weight: 19 tonnes (18.7 tons)
Armament: four 23-mm cannon with 500 rounds per gun
Performance: maximum speed 44 km/h (27 mph); range 260 km (162 miles)
Dimensions: length overall 6.54 m (21 ft 5 in); width 2.95 m (9 ft 8 in)
Powerplant: one 209-kW (280-hp) V-6R diesel engine
Users: Afghanistan, Algeria, Angola, Bulgaria, Chad, Cuba, Czechoslovakia, East Germany, Egypt, Ethiopia, Hungary, India, Iran, Iraq, Israel, Jordan, Laos, Libya, Mozambique, Nigeria, North Korea, Peru, Poland, Romania, Somali Republic, Syria, USSR, Vietnam, Yemen and Yugoslavia

The ZSU mounts the AZ-23 23-mm gun, which fires at a rate of up to 1,000 rounds per minute. It can fire while on the move at speeds of up to 25 km/h, but this reduces its accuracy by as much as 50 per cent.

Reflecting the difference in technology, the ZSU-23/4 has one extra crew member. His role is uncertain, but it is safe to assume that he acts as a loader/radio operator.

Weighing in at under 19 tons, the ZSU is surprisingly slow. Its low weight is achieved by an almost complete lack of armour. The vehicle is protected against small arms and splinters, but nothing more.

also involved in overseas sales, making use of their experience in developing state-of-the-art weapons such as the Gepard. Britain stands out among the developed nations in the lack of interest it has shown in gun-based air defence. Only the private venture Marconi Marksman has had any success in the export market.

It is only in the last 20 years that truly effective all-weather SPAAGs have been possible. With advances in electronics, their radar and sighting systems can stand up to the battering of cross-country travel. Missiles have become the dominant form of air defence since World War II, but the anti-aircraft gun remains vitally important to most armies. Given effective guidance and fire control, it is a superb weapon against low-flying threats. It also has a useful secondary ground combat capability. And when anti-aircraft weapons are mounted on vehicles, they are able to protect a mechanised force on the move.

In service with the West German, Dutch and Belgian armies, the Gepard is an outstanding weapon. It does prove to be something of a beast to camouflage, though!

 234 SWITZERLAND/ WEST GERMANY

Krauss-Maffei/Contraves 5PFZ Gepard

The **5PFZ Gepard** entered West German service in 1976 after development as a system to protect armoured formations. The type uses the hull of the Leopard 1 battle tank with reduced armour protection to carry a welded-steel turret capable of powered traverse through 360° and accommodating the weapon system.

This latter comprises two cannon, their ammunition supply and the fire control system. The cannon are located externally to avoid the problem of gun gas in the fighting compartment, and comprise two 35-mm Oerlikon-Bührle KDA weapons, which each have a cyclic rate of 550 rounds per minute firing HE incendiary, armour-piercing and semi-armour-piercing HE incendiary ammunition. The cannon have an effective range of 3000 m (3,280 yards) and, in general, fire bursts of between 20 and 40 rounds. The fire

control system is based on a computer supplied with target data by two radars: the acquisition unit has a range of 16 km (9.94 miles) and, located above the turret rear, sweeps through 360°, while the tracking unit has a range of 15 km (9.32 miles) and, located on the turret front, sweeps through a sector of 200°. Other features are optical sights, a land navigation system and an NBC protection system.

Specification
Krauss-Maffei/Contraves 5PFZ Gepard
Type: four-crew self-propelled 35-mm AA gun system
Weight: 47.3 tonnes (46.55 tons)
Armament: two 35-mm cannon with 680 rounds, and four smoke-dischargers on each side of the turret
Performance: maximum speed 65 km/h (40.5 mph); range 550 km (342 miles)
Dimensions: length overall 7.68 m (25 ft 3 in); width 3.27 m (10 ft 9 in)
Powerplant: one 619-kW (830-hp) MTU MB 838 CaM500 diesel engine
Users: Belgium, Netherlands and West Germany

The acquisition radar of Gepard has a range of 19 km, one km less than that of the ZSU. However, the Soviets are still having problems, having opted for the more complicated combined tracking and detection radar.

The radar dish on the front is the tracking radar that controls the firing of the twin 35-mm guns. Each gun fires at a rate of 550 rounds per minute, and has 320 rounds of onboard stowage.

Since the hull is the same as that of the excellent Leopard 1 tank, the Gepard has no automotive or armour problems. It can easily keep up with the formations it is tasked to defend.

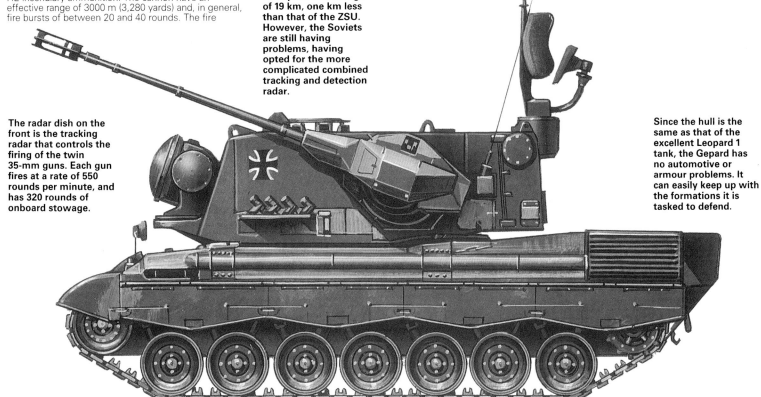

111

CHARIOTS OF FIRE

From the very first days of shooting down men in balloons to the modern radar-controlled, proximity-fused, highly sophisticated and fully automated systems, the anti-aircraft gun has always been linked to the speed and sophistication of its target.

Self-propelled anti-aircraft gun (SPAAG) systems are by no means a recent innovation. One of the first inventions was the German Panzerkraftwagen Ehrhardt, fitted with a turret-mounted 5-cm gun designed to shoot down balloons. It first appeared in 1906.

Although serious development of SPAAGs started in the 1920s, it was not until World War II that they appeared in large numbers. During this conflict such weapons were seen mounted on wheeled, half-tracked and fully-tracked mounts in a variety of calibres.

It was the Germans who led the world in SPAAG development. Although the Allies did introduce anti-aircraft systems, they were not in evidence until the later stages of the war, by which time the German air threat was not significant. In contrast, the Germans were fighting the air-land war from the first days and needed the protection SPAAGs could provide.

Most of the systems of this vintage were slow to operate, having manual controls and optical sighting systems. They ranged in calibre from 0.30 in (7.62 mm) right up to 88 mm (3.46 in) and were essentially clear-weather, daytime

World War II

The Germans were, without a doubt, the world leaders in the development and deployment of the SPAAG. The calibres grew from the 2-cm Falkvierling 38 right up to the 88-mm. They were also the first to realise the advantage of multiple gun mounts. If you use a low-calibre gun, you need to put up a lot of rounds: one of the ways to do this is to increase the number of guns firing. The Americans adopted this concept in the Maxson mount.

Introduced in 1943, the Möbelwagen was a Panzer IV hull on which four 20-mm cannon were mounted. It fired at 1,800 rounds per minute.

systems. However, given the nature of the threat, they were very much products of their time. That being said, the Israeli army still operates a SPAAG based on the old American M3 half-track. It is credited with no less than 60 per cent of the aircraft downed by land-based systems in the 1973 Arab/ Israeli war.

Post-war

In the post-war period, the Americans fielded the M42 twin 40-mm SPAAG. It was little more than a new chassis on which the older, World War II vintage, M19

turret had been mounted. Nevertheless, it was taken out and used in the Vietnam War along with the Maxson mount quad 0.5-calibre. The latter was a simple system that could be mounted on the back of a truck, and incorporated four Browning 0.5-calibre heavy machine-guns. It was quickly realised that the air threat to the US forces was insignificant. The vehicles remained, and were used in the ground role: four 0.5-calibre machine-guns make a formidable anti-personnel weapon.

The Soviets followed a very similar route. The ZSU-57/2 was the

first major Soviet SPAAG. It was a simple system incorporating two 57-mm guns in a turret. In fact, it was nothing more than the older SU-57/2 towed gun in a mechanised mount. It was replaced by the ubiquitous ZSU-23/4 in the late 1960s; this is only now being replaced by the 2S6.

Of the modern systems available, and there are numerous weapons produced as private ven-

tures around the world, perhaps the most successful has been the Kraus Maffei Gepard, based on the excellent Leopard 1 hull and mounting twin 35-mm Oerlikon KDA cannon which fire at a rate of 560 rounds per gun per minute. It is in service with the West German, Belgian and Dutch armies.

The future of the SPAAG is assured. Although the Central European threat has all but gone, the requirement for a mobile, rapid-firing, all-weather system remains in numerous troublespots around the world.

ZSU-23/4 and Krauss-M

Ammunition

The ZSU-23/4's 23-mm ammunition comes in belts of 500 rounds with one armour piercing to every three high explosive rounds. Both have a tracer base, and the AP round can penetrate 25 mm of armour at 500 metres and 19 mm at 1000 metres. Each ZSU-23/4 carries 2,000 rounds in 40 boxes of 500, and supply trucks follow about 1 km behind with another 3,000 rounds for each vehicle.

Quadruple AZP-23 23-mm cannon

Separated from the crew compartment by the armoured bulkhead, the four cannon have a cyclic rate of fire of 800-1,000 rounds per minute, per barrel. It can engage targets using one or two guns rather than all four. Targets are usually engaged with 40-round bursts.

Crew

The crew of three is made up of a driver and two turret crew: commander and gunner. Both can operate the guns and have similar gun controls. It is the commander's job to acquire and identify targets. The gunner will then take over the shoot while the commander looks for another target.

Danger: runaway gun

The ZSU-23/4's cannon sometimes continue to fire while the turret traverses after the gunner has finished a burst. This is a disagreeable experience for ground troops nearby and is one of the reasons why ZSU-23/4s travel some distance away from troops they are supporting.

Chassis

Similar to that of the PT-76 light tank, the chassis of the ZSU-23/4 has an overpressure NBC system but, surprisingly, is not amphibious. It crosses rivers on GSP ferries and can fire while afloat.

Hull armour protection

The ZSU-23/4 has only 15 mm of armour on its hull front, sloped at 55 degrees, and 15 mm of armour on the hull sides. This protects the hull from small-arms fire and shell splinters but is easily penetrated by anti-tank rockets and cannon.

Driver

The driver's cab is self-contained and he has no access to the turret. Communication is achieved by an internal intercom. The vehicle is steered by a motorbike-type arrangement, rather than sticks as found on older MBTs. Gepard has a fully automatic four-speed gearbox. It has the ability, when stationary, to execute a complete pivot: one track is driven forward, the other backwards, thus rotating the vehicle about its own axis.

SPAAGs East and West:

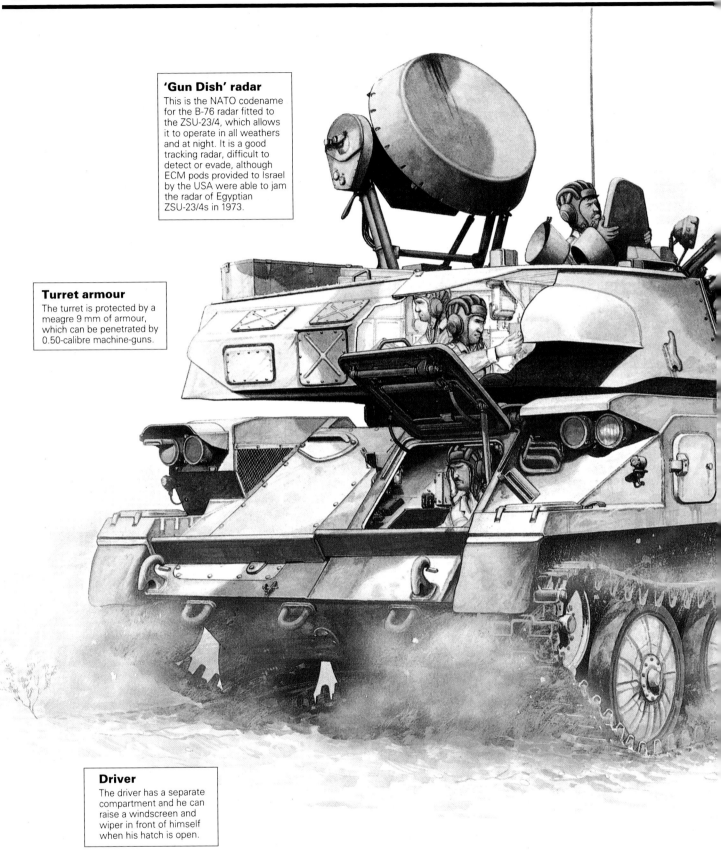

'Gun Dish' radar
This is the NATO codename for the B-76 radar fitted to the ZSU-23/4, which allows it to operate in all weathers and at night. It is a good tracking radar, difficult to detect or evade, although ECM pods provided to Israel by the USA were able to jam the radar of Egyptian ZSU-23/4s in 1973.

Turret armour
The turret is protected by a meagre 9 mm of armour, which can be penetrated by 0.50-calibre machine-guns.

Driver
The driver has a separate compartment and he can raise a windscreen and wiper in front of himself when his hatch is open.

Both ZSU-23/4 and Gepard have proved, in their own ways, to be very successful vehicles. Although Gepard has never been used in combat, unlike ZSU-23/4 which saw action in numerous Arab/Israeli wars, it has proved itself in other ways, being a popular export vehicle. ZSU relies on mass but more inaccurate firepower to achieve its aim, whereas Gepard can fire far fewer rounds but is reckoned to be much more accurate. However, as with a lot of Western technology it is far more expensive and the question must be asked, is it better to have a lot of cheap and inaccurate vehicles or a few expensive but deadly vehicles?

Far left: The Maxson mount in service at Remagen. It was a very versatile system that could be mounted on anything from a half-track down to a jeep. It was electrically powered, but aiming and guidance were entirely manual.

Firing its twin 40-mm cannon at a rate of 120 rounds per barrel per minute, the M42 'Duster' was extremely effective in an anti-personnel role, although not quite so good against aircraft.

Above: The Panzer IV chassis proved to be a useful base for a number of anti-aircraft guns. Early attempts at mounting four 20-mm cannon were dropped in favour of a single large-calibre gun – the 37-mm Flak 37. The vehicle had a crew of seven.

Vietnam War

When the Americans went into Vietnam they took with them large amounts of anti-aircraft artillery. It was quickly realised that it wasn't needed in that role, but the M42 'Duster', with its twin 40-mm cannon, and the Maxson mount quad 0.5-calibre gun were soon put to good use in the ground role. They were frequently used for perimeter defence of installations.

Above: In South East Asia, the Maxson mount emerged as a popular and successful weapon. Used largely for perimeter defence of fighting camps in Vietnam, it was an awesome weapon against unprotected infantry – as its name here suggests.

Right: Based on the M41 Walker 'Bulldog' light tank, the M42 was lightly armoured but relatively fast. However, along with the Maxsons, its most common uses in Vietnam were for convoy protection and static perimeter defence.

Modern SPAAGs

The SPAAG has always lagged behind aircraft in development, for the obvious reason that there is no point in developing a gun to counter a threat that does not exist. Modern SPAAGs have to be able to acquire, identify and engage jets that may be moving at over 500 km/h. At the same time they must also be capable of taking on slow-moving, low-flying helicopters that weave in and out of cover. Wire sights and manual traverse are things of the past; multiple radar with electro-hydraulic traverse and guns firing at up to 1,000 rounds per minute are more the order of the day.

Left: The first attempt by the Soviet Union to introduce a modern SPAAG. It was a T-54 hull with a special twin 57-mm cannon mount, and was known as the ZSU-57/2. It had a crew of five, no radar, no NBC protection and no amphibious capability.

Above: Not all SPAAGs need to be mounted on sophisticated all-terrain vehicles – it all depends on your area of operations. In this case, a West German-made Unimog truck is all that is needed for this Lebanese militiaman. The gun is the Soviet ZU-23/2, a twin 23-mm towed anti-aircraft gun.

Right: The M2 half-track and Maxson mount, as seen in World War II, is still in service today. The Israeli army has replaced the quad 0.5-calibre guns with twin 20-mm cannon and updated the vehicle, but it remains easily identifiable.

Left: With its twin Oerlikon 35-mm cannon and independent tracking and acquisition radars, the West German Gepard is an excellent weapon system. One drawback, shared with most SPAAGs, is that it is very difficult to disguise. When in operation, the rotating radar dish is extremely easy to spot.

Right: Designed to be fitted to almost any main battle tank chassis, the Marconi Marksman twin 35-mm anti-aircraft turret has been technically a successful venture. It has a highly ECM-resistant radar and twin 35-mm Oerlikon KDA cannon, as on the Gepard. The cannon fire at 550 rounds per barrel per minute and carry 230 rounds per barrel.

Search radar
Gepard, unlike its Soviet counterpart, uses two radars: one for acquisition and search, and the other for tracking. The search radar has a range of 15 km and operates in the S-band. It has a choice of six operating frequencies which are controlled by the operator. The IFF system is built in and automatic.

Engine
Gepard is powered by the same liquid-cooled 10-cylinder V-engine as is found on the Leopard MBT. It provides 610 kW (830 hp) of power, giving the Gepard a power-to-weight ratio of 18:1. The vehicle can travel at a maximum speed of 65 km/h and has a battle range of 550 km.

Hull
As Gepard shares the same hull as Leopard it has excellent cross-country ability due to the independently mounted suspension units for each road wheel. Since Gepard is not expected to be in direct combat, some of the frontal armour has been reduced to save weight.

Tony Bryan.

ffei Gepard

Optical sights

For operating closed down, the commander and operator are provided with ×1 vision ports for general observation. In addition there are optical tracking devices which provide ×1.5 or ×6 magnification. They are fully stabilised and can be used in periods when radar may not be used. They can also be used for multiple target engagements and for taking on ground targets that may try to hide in radar cover.

Crew

The crew of three is made up of a driver and two turret crew: commander and gunner. Both can operate the guns and have similar gun controls. It is the commander's job to acquire and identify targets. The gunner will then take over the shoot while the commander looks for another target.

Tracking radar

Once a target is acquired and coded hostile by the search radar, the tracking radar will pick up the target. Its function is to bring the guns onto line for the correct elevation and aim-off.

Above: Italy's very successful and revolutionary weapon, the OTO-Melara Otomatic. It can fire its 76-mm gun, in five- or six-round bursts, out to a range of over 4000 metres against aircraft. Its main strength, however, is that it can engage helicopters at 6 km – before they have a chance to launch anti-tank missiles. It is probably the prototype for future SPAAGs.

Left: As a model in their VAB range, the French offer the VDAA (Véhicule d'Auto-Défense Antiaérienne). It has twin 20-mm cannon and a sophisticated pulse-Doppler radar system.

In the early 1970s, the US Army decided it needed a replacement for its Vulcan system. The main contender was the DIVAD (Divisional Air Defence) system, known as Sergeant York.

Above: Sergeant York, which bore an uncanny resemblance to the ZSU-57/2, was a disaster from beginning to end, and testing was discontinued at the end of 1984. Vulcan is still in service.

Sergeant York fiasco

In the 1970s, the US Army decided that it needed to improve its air defences. After a protracted competition, the Ford Aerospace Sergeant York was chosen as the M247 DIVAD, or Divisional Air Defence system. The M247 was an object lesson in how not to develop a new weapon. A pair of well-proven Bofors 40-mm cannon was to be mated to a fire control radar developed from that of the F-16 fighter, with the whole thing mounted on a modified M48 tank chassis.

Nobody seemed to have considered that the integration of systems from such widely varied sources might prove difficult. Similarly, the choice of an M48 chassis ensured that the Sergeant York would not be able to keep up with the M1 Abrams tanks and M2/3 Bradley fighting vehicles then about to enter service.

From the start, the M247 caused problems. In one test, Sergeant York is reported to have demolished a portable toilet, its fire control radar having mistaken an extractor fan for the rotor blades of a helicopter. Fortunately, nobody was 'engaged' at the time! In 1985, when it became clear that even after the expenditure of more than a billion dollars the system would never work as advertised, the whole programme was cancelled.

SINAI SAM

In 1973 the Israeli air force was caught completely by surprise as it tried to repeat the successes of six years earlier. This time the Egyptians were ready. From the long-range SA-6 'Gainful' down to the close-in hail of fire from the ZSU, the Egyptian air defence was almost impenetrable. Almost.

Above: The Israeli F-4 Phantom was the mainstay of the Israeli air force throughout the numerous Arab wars, with the exception of the campaign in Beka'a in 1982. Fitted with ECM pods to jam and confuse the Arabs' large, radar-guided strategic and tactical missiles, and with chaff and flare dispensers to decoy their heat-seeking missiles such as the SA-7, the Phantom was a difficult target.

"Planes starred with blue hexagons plummeted towards earth as if they were falling comets," an Egyptian general reported gleefully after watching the Israeli air force fail to knock out his Soviet-made PMP pontoon bridges across the Suez Canal on the first day of the 1973 Yom Kippur War.

In a desperate bid to hold the surprise Egyptian advance into Sinai, hundreds of Israeli Phantoms and Skyhawk fighter bombers were sent to take out the vital bridges on 6 October. A storm of surface-to-air missiles (SAMs) and fire from anti-aircraft guns met them.

Three hours into the battle, the once all-powerful Israeli air force turned tail for home after 13 planes had been blown out of the sky. More importantly, none of

the pontoon bridges were put out of action, allowing the Egyptian army to consolidate its bridgehead in Sinai unmolested by Israeli air power.

Jamming fails

Only six years earlier, Israeli planes had roamed at will over Egypt and destroyed the Arab nation's air force in a dramatic first strike against its air bases during the Six Day War. Realising that they could never match the skill of Israeli fighter pilots, the Egyptians built up a massive air defence system, with the latest Soviet-made SAMs, anti-aircraft artillery (AAA) guns and radars.

The intensity of the air defence system took the Israelis completely by surprise. Their jamming equipment was unable to counter the new missiles, radars

Above: The triple rockets of the SA-6 'Gainful' launcher. In the 1973 Yom Kippur War, the Israelis were taken by surprise by this weapon.

Right: Captured SA-2 missiles on the western bank of the Suez Canal. Once the Israelis crossed the canal, the Egyptian air defence umbrella folded completely.

Below: An Egyptian missile battery on the west of the canal. Amazingly enough, the sites – minus the missiles – are still visible today.

and communications links being used in action for the first time by the Egyptians.

American advice and their own experience against old model SA-2s and SA-3s made the Israelis confident that low-level approaches and electronic countermeasures (ECM) equipment, which jammed guidance links from ground control stations, would keep them safe. They got a big shock on that October afternoon.

Coming in low over the Sinai, the Israeli pilots saw a mass of small red dots rising to meet them. These were SA-7 heat-seeking missiles fired by Egyptian infantrymen in forward trenches with hand-held launchers. As the Phantoms and Skyhawks powered towards the Suez Canal, many of the SA-7s couldn't keep up with

the fast jets and ran out of fuel. A few got good locks on the Israeli's red-hot jet exhausts and found their mark.

With the canal on the horizon, a curtain of fire arose from ZSU-23/4 radar-guided mobile AAA guns. The air was full of red tracer and the Israeli pilots immediately pulled their aircraft up to escape this furious barrage. Warning alarms started blaring to alert the pilots of radar locks from SAM systems and missile launches were spotted from the far bank of the canal.

Jamming pods were activated but the missiles kept coming – the Israelis had met the SA-6 'Gainful' for the first time. Onboard semi-active homing guidance systems made the missile immune to jamming – violent high-*g* evasive manoeuvres were the only way to get away from the SA-6s.

When the Israeli pilots dived to zero altitude to escape the new missiles, ZSU-23/4s scored more hits on the fleeing Phantoms and Skyhawks.

Three hours into the attack, Israeli air force commander General Benjamin Peled ordered his aircraft to keep 15 kilometres to the east of the Suez Canal. The Egyptians were jubilant when they monitored the message being sent in clear by a panicked Israeli high command. The disaster of 1967 had been avenged.

Victory over Beka'a

Nine years later, when the Israelis went into battle against Syria's SAM defences in Lebanon, the odds were definitely loaded in their favour.

The Israeli air force commander, Major General David

Egyptian air defence umbrella

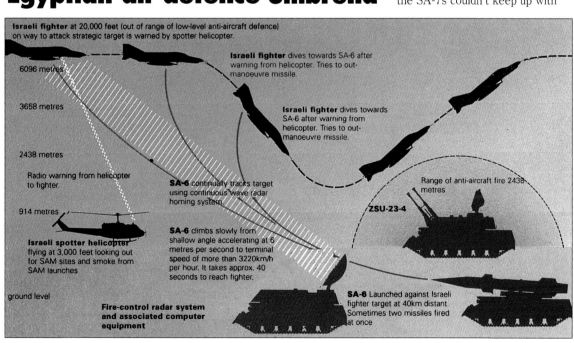

Israeli fighter at 20,000 feet (out of range of low-level anti-aircraft defence) on way to attack strategic target is warned by spotter helicopter.

6096 metres

3658 metres

2438 metres

914 metres

ground level

Israeli fighter dives towards SA-6 after warning from helicopter. Tries to out-manoeuvre missile.

Israeli fighter dives towards SA-6 after warning from helicopter. Tries to out-manoeuvre missile.

Radio warning from helicopter to fighter.

Israeli spotter helicopter flying at 3,000 feet looking out for SAM sites and smoke from SAM launches

SA-6 continually tracks target using continuous wave radar homing system.

SA-6 climbs slowly from shallow angle accelerating at 6 metres per second to terminal speed of more than 3220km/h per hour. It takes approx. 40 seconds to reach fighter.

Fire-control radar system and associated computer equipment

Range of anti-aircraft fire 2438 metres

ZSU-23-4

SA-6 Launched against Israeli fighter target at 40km distant. Sometimes two missiles fired at once

Left: The F-15 ruled the skies over the Beka'a Valley, claiming 40 kills for no loss. The Israelis were able to amass such high scores because of their superior tactics, tight control and effective early warning.

Below: Remotely piloted vehicles (RPVs) were used widely throughout the Beka'a Valley, for both decoy and recce work. This is an IAI Scout, complete with TV camera and other gear. The cedar tree symbols on the side denote 14 successful missions over Lebanon.

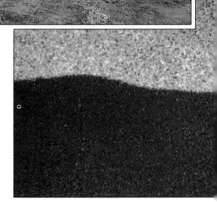

Specially-modified Boeing 707s provide stand-off electronic countermeasures for the Israeli air operations. As well as carrying an extensive ELINT suite, the 707s carry out jamming and transmit spoof signals

Israeli SAM tactics

Central to the Israeli victory in the air over Lebanon was the destruction of the Syrian SAM defences. Operating under strict control of Hawkeye command and control aircraft, specially-equipped Phantoms knocked out the defending radars piecemeal, leaving the helpless missiles and support vehicles at the mercy of the armed-to-the-teeth Kfir bombers. The quality of mercy may not be strained, but that wasn't what was dropped from heaven above. The Kfirs were relentless. The diagram below shows how a typical SAM strike was carried out, from the introduction of the RPVs to the eventual destruction of the target by the Kfirs.

F-15 Eagles provide fighter cover for the entire strike. These fly over the battlefield, and their priority is protection of the Hawkeye

Above: The Israelis did not have it all their own way. This F-4 was hit by Syrian missile fire. But the Israelis achieved a massive victory. On 9 June 1982 they took out 19 SAM batteries and 29 MiGs for no loss. By the end of the campaign, more than 90 aircraft and 30 SAM batteries had been lost by the Syrians, compared with 13 Israeli aircraft and helicopters.

The **E-2 Hawkeye** is the linchpin of the operation, providing airborne radar coverage and keeping in contact and controlling all the facets of the strike

1b The RPV is first detected by search radar. Thinking the RPV is an attacking aircraft, the Syrians turn on the SAM tracking radar, located next to the missiles

1a Mastiff RPVs are sent over towards the SAM sites to trick the Syrians into turning on the tracking radar of the SAM system. These RPVs are remotely flown from the Hawkeye and have radar signatures similar to an attacking aircraft

2b Once in the vicinity of the SAM site, the F-4 sets up a racetrack at low level, using the cover of hills to shield it from the Syrian radar

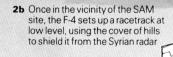

2a McDonnell Douglas F-4E Phantoms armed with Shrike or Standard anti-radiation missiles (or both) are on station. When the RPV approaches the suspected SAM site, the Hawkeye calls them into the a

Ivri, described the destruction of the Syrian SAM system as the "climax of the war", and put victory down to long practice and combined arms teamwork, which turned the strike force into a "well-oiled fighting machine."

For years a massive effort was put into analysing Soviet-made SAMs, their control radars, command systems and tactics. Never again would the Israeli air force be unprepared.

Operation Peace for Galilee

began on 6 June with massive Israeli armoured attacks on PLO bases throughout southern Lebanon. Syrian troops massed in the Beka'a Valley were soon drawn into the fighting, and the Israelis decided to eliminate the Arab air defence system.

Syrian SAMs defeated

At 1400 on 9 June the Israeli air force went into action over the Beka'a Valley. Israeli-made Scout and Mastiff remotely piloted vehicles (RPVs), or drones, were sent up to simulate attack aircraft and goaded the Syrians into switching on their SAM radars. Soon ECM warfare officers in Israeli Boeing 707s and Hawkeye AWACS radar planes were almost overwhelmed with data on Syrian radar frequencies and pinpoint fixes on SAM sites. Powerful jamming equipment now blinded the SAM's radars, and target information was quickly passed to attack units.

First, Israeli artillery and commando teams blasted the

Syrians' AAA positions, providing low-level protection for the SAM batteries. The 188 Phantoms, F-16s, F-15s and Kfirs were launched in two waves to take out the SAM sites with anti-radiation Shrike missiles, which home in on radar signals, cluster bombs, 'smart' bombs and conventional iron bombs. To further confuse the Syrians, the Israelis operated their aircraft in small groups and attacked from multiple directions. The SAM crews didn't know where the next attack was coming from. In a futile attempt to protect their missiles the Syrians tried to hide them with smokescreens, but this only helped the attackers find their targets.

It was all over in two hours. Seventeen of the 19 Syrian SAM batteries lay in ruins, the remaining two batteries were damaged and most of the Arab radar network was smashed. Not a single missile had left the ground to challenge the Israeli bombers.

As Israeli aircraft were finishing off the SAM sites, the Syrians

threw their air force into the fray to protect what was left of their SAM system. Israeli Hawkeyes gave ample warning of the Syrian attack and F-15 Eagles flying top cover for the bombers were vectored to intercept. In a few minutes 29 MiGs were shot out of the sky, and in the following days the Israelis would push their score to 95 MiGs downed. With good planning and intelligence, superb electronic equipment, all-arms co-operation and determined flying, the Israelis had broken down the Syrian SAM system into its component parts and eliminated them one by one.

Major General Ivri was very pleased with his boys' performance, saying, "The IAF has indeed gained a brilliant victory over the skies of Lebanon. Moreover, far from sitting back on its laurels, the commanders are already hard at work; nothing is left to chance – the next conflict is soberly assessed and new ideas are taking shape." The Israelis had learned the lessons of 1973.

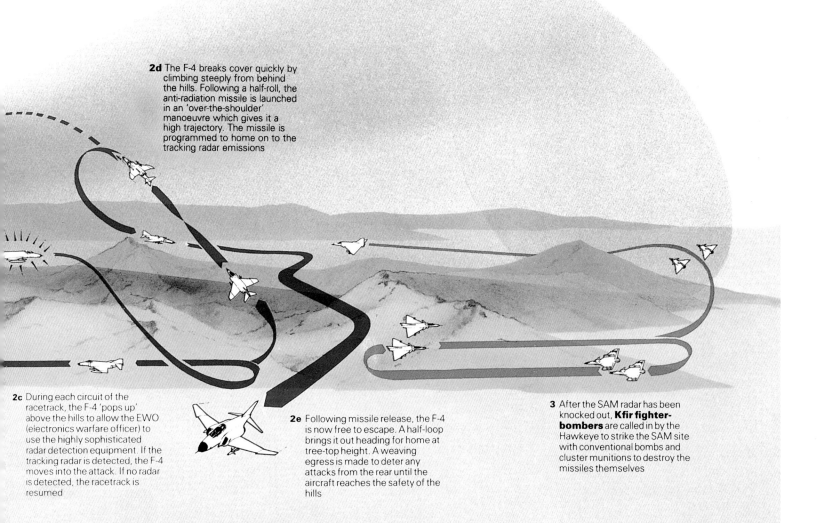

2d The F-4 breaks cover quickly by climbing steeply from behind the hills. Following a half-roll, the anti-radiation missile is launched in an 'over-the-shoulder' manoeuvre which gives it a high trajectory. The missile is programmed to home on to the tracking radar emissions

2c During each circuit of the racetrack, the F-4 'pops up' above the hills to allow the EWO (electronics warfare officer) to use the highly sophisticated radar detection equipment. If the tracking radar is detected, the F-4 moves into the attack. If no radar is detected, the racetrack is resumed

2e Following missile release, the F-4 is now free to escape. A half-loop brings it out heading for home at tree-top height. A weaving egress is made to deter any attacks from the rear until the aircraft reaches the safety of the hills

3 After the SAM radar has been knocked out, **Kfir fighter-bombers** are called in by the Hawkeye to strike the SAM site with conventional bombs and cluster munitions to destroy the missiles themselves

THE DEADLY ZSU

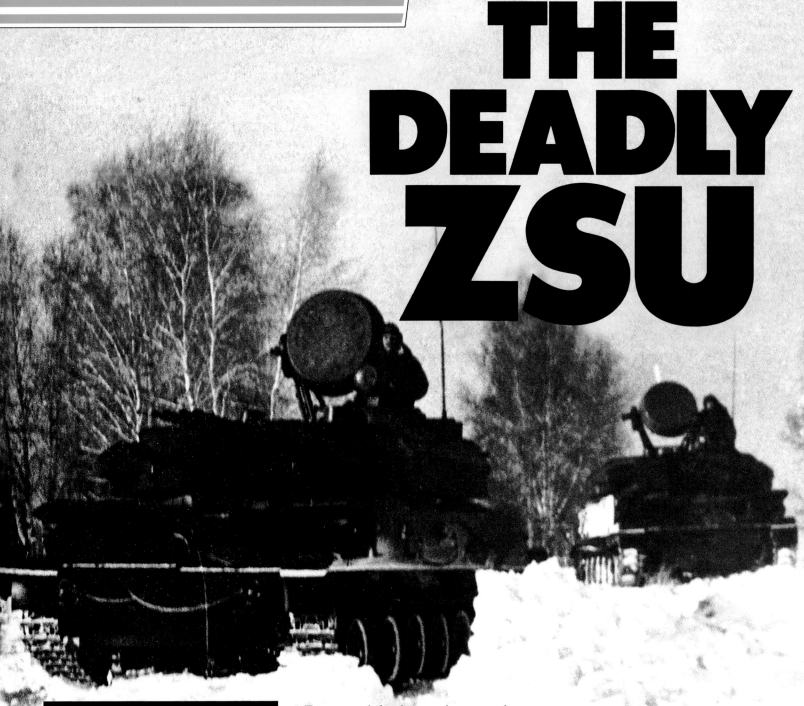

The ZSU-23/4 is one of the most effective anti-aircraft artillery pieces in large-scale service today. Firing at an incredible 1,000 rounds per barrel per minute, it can pump over 750kg of high explosive into the sky in 60 seconds. It is rightly a much-feared weapon.

As one of the few armies properly to appreciate the importance of anti-aircraft fire, the Soviet Army introduced the four-barrelled, lightly-armoured ZSU-23/4 in the late 1960s. It saw active service in the 1973 Arab/Israeli war, in which the Israeli air force reckoned that 30 of their 80 combat losses in the first three days of the war were caused by ZSU fire.

The four-barrelled turret is fitted with the AZP-23 cannon, which fires its 23-mm shell (either armour-piercing incendiary, high-explosive incendiary or frag) to a maximum range of 7000 metres, although its effective AA range is more on the order of 3000 metres.

The chassis of the vehicle is based on the PT-76 light tank and, surprisingly, given Soviet-era doctrine, it is not amphibious,

Operating in pairs to provide mutual support is the usual deployment for the ZSU. This also ensures that cover is achieved if one vehicle fails.

but it has and NBC facility as well as crude night-vision devices. It is manned by a crew of four – three turret crew and the driver.

As with all Soviet-era equipment, there has been a constant improvement programme. The original ZSU had an analogue computer and problems with its radar dish. This analogue computer has now been updated and replaced by a more sophisticated and modern digital computer. Likewise, the radar system has been upgraded so that it can operate at different frequencies and more easily pick up moving targets that are hidden in ground clutter.

1 Deploying the ZSU

The ZSU will always operate in pairs, and often in fours. Given the high rate of fire and the rather meagre onboard ammo stowage (if it could fire at its maximum rate, it would fire all its ammo in just 30 seconds), a ZSU unit will always be followed by unarmoured resupply vehicles. The exact deployment of the vehicles will be up to the platoon commander, who will travel in a BTR-60 PU, but his usual mission is to protect the command vehicles. In some circumstances, one pair may be sent forward to provide cover for a forward detachment, or the lead battalion of the motor rifle regiment if the air threat is particularly great. During river crossings the ZSUs will be off at the flanks on the river banks some four to five kilometres from the crossing site. For a pilot, a river is an ideal flight path that will take him onto his target; the ZSU's deployment both up- and downstream of the site will disrupt this route.

Flight path suppression involves denying a flight path to an enemy pilot. In this case, the river would provide him with an ideal route since it leads straight onto the target. By positioning ZSUs up- and downstream of the crossing site, all approaches can be covered.

Flight path supression

ZSU 23/4 operate in pairs to cover all approaches to the crossing site made along the river. It prevents pilots from using the river as a guide to take them onto the target.

cross river to provide cover from far bank

crossing site

pair ZSU 23/4

Acquisition, lock-on, in range

target acquired

target locked-on

target in range

20 km

8 km

3 km

2 Engagement sequences

There are three methods of engagement: radar, electro-optical and optical. The first is the primary mode. The gun dish will normally be set on automatic surveillance or sector scan, the sector being decided by data from other radar systems. Once the gun-dish radar picks up a target, at about 20 kilometres out, it will interrogate it to establish if it is friendly, using its IFF (Identification Friend or Foe) sensor. If the contact does not give the correct response, the target will be coded as hostile, in which case the computer will initiate the tracking radar at about an eight-kilometre range. The co-ordinates will automatically be fed into the computer to establish the correct elevation and aim-off. The tracking computer will then attempt to get a lock-on. This takes about 20 seconds. A modern jet will cover the eight kilometres in about 30 seconds, so the target should just come into range as the computer locks on.

Operating on its own, i.e. not linked to other radar systems, the ZSU-23/4 can acquire targets up to 20 km distant. It will continue to track the target and achieve lock-on at 8 km. In the time the computer takes to calculate the necessary angles, the target will be in weapons range.

---

3 Low-level engagements

Should the target come in low and under the tracking radar, or if the computer fails to get a lock-on, perhaps because of electronic countermeasures, ZSU can fire in an electro-optical mode. The radar will still acquire the target but cannot lock on. In this case, the range data from the computer will be read by the gunner and applied to his optical sight. The computer will still be able to tell the guns the correct elevation, and it is up to the skill of the gunner to bring the guns to bear. This may be possible when shooting at a slow-moving helicopter, but to engage a jet flying across your front at 1000 metres you would have just two seconds to track it and fire the gun. But if you managed this, you could get off 140 rounds in that time.

The only way to defeat the radar is to use the folds of the earth and come up inside the acquisition range. The computer does not have time to lock on, but can get a range to the target.

No lock-on

Using the cover of the ground, the plane can come in under cover. Lock-on will not occur but ranging radar will return the distance to the target.

incoming attacker

radar dead space

under 8 km

4 ZSU versus ground targets

The final method of firing the gun is in the optical mode. This is the fastest method of engagement, but is massively inaccurate. With no range information, it would be highly wasteful of ammunition. In reality, this tactic is not possible against any but the slowest-moving helicopter; nevertheless it is viable against ground targets. The US Army has estimated that against static ground targets ZSU has a 97 per cent chance of hitting at 100 metres, falling off to only a 31 per cent chance at 1500 metres. Not particularly impressive, given that a battle tank has an 80 per cent chance of a first-round hit at 3000 metres, rising to a 97 per cent chance with three rounds.

Since it can put such a heavy weight of fire into a target, the ZSU makes an ideal anti-light armour weapon. However, it is wildly inaccurate – the radar is inoperative, and ranging and fire control are manual.

Direct fire

enemy APC

1000 m

If the ZSU is ambushed or surprised, it can fire in the direct role. It is unguided and no ranging devices are used.

The Russian armed forces still have the capability to erect a comprehensive air-defence umbrella. The ZSU-23/4 is the final spoke of that cover, with its short range and high rate of fire.

5 The air defence umbrella

ZSU does not operate on its own, but is just one part of the umbrella of the Russian air defence. The umbrella is conceived to provide air cover against all types of threats, at all heights and ranges. Indeed, ZSU is one of the last measures in that cover, providing point rather than area defence: i.e. it is tasked to protect an individual or group of vehicles, not an area of ground. This was exploited by the Arabs against Israeli planes. The Israeli jets were forced to fly low to avoid the enemy's SA-6 surface-to-air missiles, and as they did so they flew into the weapons envelope of the ZSUs. In contrast, British Army large-scale point defence consists of a machine-gun firing 800 7.62-mm rounds per minute (rather than 4,000 23-mm rounds) mounted on a stick. This must inspire great fear in the hearts of enemy pilots.

Left: The SA-6 'Gainful' is one of the longer-range tactical weapons in the Russian air defence umbrella. It has a maximum range of 24km.

Above: the main Russian air defence weapon is the SA-8 'Gecko'. It can engage aircraft at an altitude of 13km.

Artillery has dominated 20th-century battlefields from the jungle to the Arctic. Here, the power of a US Army 175-mm gun is unleashed at the height of the Vietnam War.

SELF-PROPELLED ARTILLERY

With calibres from 105 mm right up to 203 mm and the ability to pour down tons of high explosive on targets miles away – targets the guns never see – self-propelled artillery is the battle-winning weapon of today.

0620 hrs. Captain Hendrie Bruce stood on the heaving landing craft among the squat grey shapes of the Priest self-propelled artillery guns. Through the spray and spume the pale light of dawn slowly revealed the dark outline of the coast ahead.

0640 hrs. Bruce ordered ranging salvoes to be fired at the beaches. The waiting gunners leapt into action. The 105-mm guns roared and bucked and the landing craft wallowed as the shells ripped through the wind towards the far coast.

0643 hrs. The first shells burst on the beach and the smoke drifted quickly out to sea.

0645 hrs. The divisional artillery – three regiments of SP guns – opened fire on the target, codename COD.

"Great clouds of smoke, dust and brick dust rise from the target area and blow out to sea, obscuring target and covering sea and ships in fog of war . . . The enemy, stung into activity, begins to pour a heavy fire from guns and mortars onto the beaches and among the craft approaching . . ."

It was 6 June 1944 – D-Day – and 9 (Irish) Field Battery Royal Artillery (now 24 (Irish) Battery) had just fired the first shells of the invasion to hit the enemy coast. It was coming back to avenge the last chaotic and desperate battles it had fought at Dunkirk in 1940.

As part of British 3rd Division the Battery landed on Queen Beach at Hermanville-sur-Mer – target COD. As the ramps went down

German lash-ups of World War II

At the outbreak of World War II, German ingenuity was severely tested to produce the rather unlikely combination of a French Lorraine chassis and the German SIG 33 15-cm gun.

In many ways the German army's heady successes of 1940 and 1941 were a long-term headache for the nation's war planners. They had not envisaged a European war breaking out before 1943 at the earliest, and both the German armed forces and the defence industry were badly placed when war broke out in 1939.

One of the consequences of this was the shortage of self-propelled artillery. In 1940 most, if not all, German artillery was towed, some of it by horses. This was a terrible weakness in the execution of Blitzkrieg, the German tactical concept.

As a result, a number of 'lash-ups' were put together. By and large, they consisted of taking an existing gun and sticking it onto whatever chassis they had a lot of, one of the most successful being the unholy marriage of the captured French Lorraine chassis with the SiG 33 15-cm gun. So successful was it that it was adopted as a standard gun and put into general service!

the Priests rumbled out of the landing craft and through the waves, some firing with the surf washing about their tracks, and then up the beach. 9 Battery was soon in action just outside the town, and that afternoon its battle started in earnest as 21st Panzer Division tried to smash the British back into the sea.

Allied punch-through

They remained in the same gun position for 32 days as the breakout battle stormed around them – administering the 'last rites' with their Priests whenever the infantry or tanks called for help. Finally, the Allies punched through and the long drive to Germany began.

The guns with which the British opened the 2nd Front that fateful day were very different from those of 1940 or even of the German defenders. Instead of being a cannon mounted

SELF-PROPELLED ARTILLERY Reference File

205

UNITED KINGDOM 🇬🇧

Vickers AS-90

A potent new weapons system, the **AS-90** (Artillery System for the 1990s) is a self-propelled 155-mm (6.1in) howitzer resulting from Vickers' belief that a high-quality SP artillery equipment requires the use of a custom-designed rather than converted main battle tank chassis, and also construction on a modular basis for ease of maintenance and updating. It was thought that the main market for such an equipment would lie outside the UK, so Vickers teamed with the British branch of Cummins, an American company, and Verolme, a Brazilian concern. These are responsible for the powerpack and chassis respectively.

Built throughout of welded steel, the AS-90 appeared in prototype form during 1986. The main ordnance has a 39-calibre barrel, and the use of an hydraulically powered ramming system with a ready-use supply of 40 rounds allows a burst rate of three rounds in 10 seconds, or six rounds per minute for a short period, or two rounds per minute for a sustained period. The ordnance fires the full range of NATO 155-mm (6.1-in) ammunition, typical range for the HE projectile being 24700 m (27,010 yards) and for the extended-range full-bore projectile 32000 m (34,995 yards). The AS-90 has an advanced fire-control system.

Specification
Vickers AS-90
Type: four-crew self-propelled 155-mm

(6.1-in) howitzer
Weight: 36 tonnes (35.43 tons)
Armament: one 155-mm (6.1-in) howitzer with 40 ready-use rounds on a mounting capable of elevation between −5° and +70° in a powered turret with 360° traverse, and one 7.62-mm (0.3-in) AA machine-gun with 500 rounds

Performance: maximum speed 55 km/h (34 mph); range 350 km (217 miles)
Dimensions: length overall 9.7 m (31 ft 10 in); width 3.3 m (10 ft 10 in)
Powerplant: one 447-kW (600-hp) Cummins VTA 903T-600 diesel engine
User: UK

Above: The Soviet 2S3 152-mm gun is technologically rather straightforward, but none the worse for that. It roughly equates in terms of performance to the US Army's M109 155-mm gun.

Left: The mighty M110 203-mm gun is nuclear, chemical and conventional HE capable, and fires rounds out to an incredible 30 km. Unfortunately, the British Army has a rather meagre 16 of these.

The only ex-Warsaw Pact country to produce its own artillery was Czechoslovakia with this 152-mm DANA. The gun is the same as that of the Soviet 2S3 and it is mounted on a Tatra 815 8×8 truck. As something of a hybrid, the DANA was not a howling success but the Czechs did manage to persuade Libya to purchase the system.

on two wheels and drawn by a lorry or horses – something that Henry VIII would have recognised – these artillery guns looked more like tanks. The gun or howitzer, designed to fire long distances at a target unseen by the gun crew, was permanently attached to a tracked and armoured carriage – often an old tank chassis. These self-propelled guns were much more mobile and flexible than towed artillery and provided the gun crew with more protection. Many of the German guns on the beaches at D-Day were still horse-drawn and stood no chance of escape once the bombardment had started.

Growing need

As modern armies have become more mechanised and armoured and as the pace of battle increases, so has the need for SP artillery become more important. Today most armies rely on these guns, except for airborne and marine operations where lighter equipment is needed for air portability.

Self-propelled firepower can be awesome – in the run-in on D-Day 200 guns fired 18,000 rounds, each weighing about 35 lb. Most SPs in NATO today are 155-mm and fire a shell weighing nearly three times that of the World War II guns – the storm of fire that they can put down is hard to imagine. They don't fire only high explosive; their armoury includes parachute flares for illumination, smoke to

206

USA

Cleveland Army Tank Plant/ Bowen-McLaughlin-York M109

The **M109** is a self-propelled 155-mm (6.1-in) howitzer, and in numerical terms the most important artillery equipment in current Western service. It was developed from the late 1950s as a replacement for the M44 with greater operational capabilities and better crew protection.

The first variant had an ordnance with a 20-calibre barrel and, stabilised by two rear-mounted spades, could fire its 28 ready-use projectiles (HE, submunition dispenser, 1-kiloton nuclear, chemical, Copperhead laser-guided anti-tank, illuminating and smoke) to a typical range of 14630 m (16,000 yards) or further when using

the rocket-assisted HE type. The **M109A1** introduced a longer 33-calibre barrel to boost projectile range to a standard 18105 m (19,800 yards) or rocket-assisted 24005 m (26,250 yards). The **M109A2** has several ordnance improvements, the **M109A3** is any earlier vehicle brought up to M109A2 standard, the **M109A4** is any earlier vehicle upgraded with an NBC system and a package of reliability improvements, and the **M109A5** is another considerable improvement.

Specification
Cleveland Army Tank Plant/BMY M109A2

Type: six-crew self-propelled 155-mm (6.1-in) howitzer
Weight: 24.948 tonnes (24.55 tons)
Armament: one 155-mm (6.1-in) howitzer with 36 ready-use rounds on a mounting capable of elevation between −3° and +75° in a powered turret with 360° traverse

Performance: maximum speed 56 km/h (35 mph); range 346 km (215 miles)
Dimensions: length overall 9.12 m (29 ft 1.5 in); width 3.15 m (10 ft 4 in)
Powerplant: one 302-kW (405-hp) Detroit Diesel 8V-71T diesel engine
Users: 25 countries worldwide

Ranges of the guns

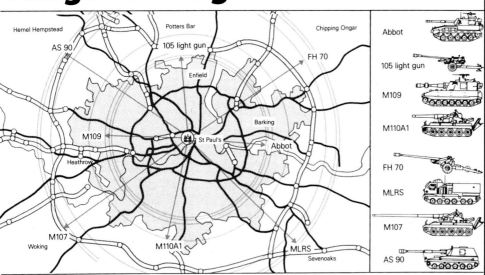

The ranges of these weapons are shown superimposed on a map of London, with a gun located at St Paul's Cathedral. MLRS (Multiple Launch Rocket System), M107 and AS-90 can all reach beyond the M25 and could take out Heathrow. Even the smallest-calibre weapons, the 105-mm Abbot and Light Gun, could engage most of London.

With guns firing at such ranges, accuracy is very important. Such things as weather, wind speed at both the target end and the gun end, and even how much the Earth rotates in the time the round is in the air must be allowed for. Gunners should be able to get the first round within 400 metres of the intended target.

cover movement, and even coloured marker shells to help direct bombers onto a target. Some of these big guns can also fire nuclear shells, but these may be phased out quite soon.

Earth's rotation

But artillery guns are not pinpoint destruction weapons. The inevitable inaccuracies that come from firing 10, 20 or even 30 km – from wind and even the Earth's rotation – mean that you cannot rely on them to knock out bunkers or tanks. But the blast, terror and surprise of a sudden heavy 'stonk' often leaves the defenders too stunned to move for several minutes. Because of this effect, attacking tanks and infantry try to get onto a target almost before their artillery has finished, risking casualties from their own guns in order to catch the enemy still too shocked to fight.

Some of the earliest SP guns were used in World War I to attack balloons and aircraft. The British simply mounted light, quick-firing guns on lorries that could almost chase the slow, ponderous aeroplanes around the countryside, Keystone Cops style!

Experiments continued after World War I but it was the Germans – faced with fluid, fast-

---

207
FORMER USSR
SO-122 Gvozdika (2S1)

The basic weapon of the motor rifle and tank regiments of the Soviet army is the **SO-122 Gvozdika** (Carnation), or **2S1**. The gun is a modified version of the towed D-30, the standard Soviet towed gun, which has a depression from –3 degrees to +70 degrees, giving it a maximum range of 15.3 km. It can fire chemical, illuminating, smoke and conventional HE as well as a self-defence HEAT anti-tank round. This can penetrate 460-mm of armour at a range of 1000 metres. The normal ammunition load is 40 rounds: 32 HE, six smoke and two HEAT. A power-rammer allows the gun to fire at a sustained rate of five rounds per minute.

The layout of 2S1 is similar to that of the M109, with the engine,

transmission and drive at the front and a fully enclosed turret at the rear. Standard equipment includes an NBC filtration system and full night vision devices for the driver and commander. The vehicle is fully amphibious.

The chassis of the 2S1 is used for a number of other armoured command and reconnaissance vehicles (ACRVs) fitted with the 'Big Fred' artillery/mortar locating radar. They are to be found attached to 2S1 batteries.

Specification
SO-122 Gvozdika
Type: four-crew self-propelled 122-mm (4.8-in) gun
Weight: 16 tonnes (15.75 tons)

Armament: one 122-mm (4.8-in) gun
Performance maximum road speed 60 km/h (37 mph); range 500 km (310 miles)
Dimensions: length 7.30 m (23 ft 11½ in); width 2.85 m (9 ft 4 in); height 2.40 m (7 ft 10½ in)
Powerplant: one 179-kW (240-hp)

YaMZ-238V V-8 water-cooled diesel
Users: Russia and 13 other countries worldwide

208
FORMER USSR
SO-152 Akatsiya (2S3)

Originally called the **M-1973** by Western intelligence, this self-propelled 152-mm (6-in) gun/howitzer is known in Russia by the service and production designations **SO-152 Akatsiya** (acacia) and **2S3** respectively. The equipment entered service in 1971 as replacement for the M-46 gun in tank, motorised rifle and artillery divisions. The SO-152 is based on the chassis of the SA-4 'Ganef' SAM system revised with six rather than seven road wheels.

The powered turret is carried at the back of the vehicle over the rearmost four road wheels, and features a powerful ordnance based on that of the D-20 towed gun/howitzer with a 34-calibre barrel. It is possible that an automatic loader is fitted, and it is

assumed that this battlefield vehicle is fitted with an NBC system as well as the standard night vision equipment. The weapon can fire the usual range of Russian projectiles in its calibre: the HE, smoke, illuminating and 2-kiloton tactical nuclear types are delivered to a range of about 18500 m (20,230 yards), and the rocket-assisted HE type to 24000 m (26,245 yards); there are also kinetic- and chemical-energy projectiles for use at short range against tanks.

Specification
SO-152 Akatsiya
Type: six-crew self-propelled 152-mm (6-in) gun/howitzer
Weight: 23 tonnes (22.64 tons)
Armament: one 152-mm (6-in) gun/

howitzer with 46 ready-use rounds on a mounting capable of elevation between –3° and +65° in a powered turret with 360° traverse, and one 7.62-mm (0.3-in) AA machine-gun with 1,000 rounds
Performance: maximum speed 55 km/h (34 mph); range 300 km (186 miles)

Dimensions: length overall 8.4 m (27 ft 6.75 in); width 3.2 m (10 ft 6 in)
Powerplant: one 388-kW (520-hp) D-12A diesel engine
Users: East Germany, Iraq, Libya and USSR

moving armoured battles in the mud of the Russian steppes in 1941-2 – who were the first to address the problem seriously. Their answer was to mount field guns on captured or obsolete tank hulls. The Wespe and Hummel (Wasp and Bumble-bee) were the result of this, but sometimes the gunners would create their own versions with whatever they could find locally.

The British were quick to follow suit, but their first – the Bishop – was not a great success and was replaced by the American Priest, named after the tall 'pulpit' for its anti-aircraft machine-gun. This reliable weapon was the forerunner of the modern American family of SP guns that are used by most NATO forces today.

The M109 is probably the most common SP in NATO, and is used by the British and Germans as well as the Americans and many other nations. It is a huge beast, mounting a 155-mm gun that can fire a 95-lb shell out to 18 km. The turret is mounted at the back of the vehicle so that ammunition can be passed in easily. It also carries shells in the turret, but these are usually reserved for emergency 'crash actions' and anti-tank engagements – a direct hit with one of their shells would probably take the turret clean off most light tanks!

Left: The amazing South African G6 155-mm gun can fire a 45-kg round over 24 miles! It was designed by the late Dr Bull, the inventor of the so-called Iraqi supergun.

Below: Although it has now been replaced by the GIAT GCT, the Mk F3 was for a long time the mainstay of the heavy artillery of the French army. It was a popular export weapon in the Middle East and South America.

209
SO-203 (2S7)

FORMER USSR

Given the NATO designation **M-1975** on its first appearance, the Russian weapon otherwise known by its service **SO-203** and **2S7** is a 203-mm (8-in) gun, and as such has been one of the world's most powerful items of self-propelled ordnance since it entered service in 1975.

As the long barrel is designed for the delivery of projectiles over very long range, the ordnance is located on an exposed mounting at the rear of the substantial chassis, but enclosed accommodation for two or four men is provided in a cab at the front of the hull. The ordnance has a simple powered loading device operated by a loader in a 'crow's nest' at the left-

hand rear corner of the vehicle. A burst rate of two rounds per minute and a sustained fire rate of one round every two minutes have been quoted.
The 2S7 can fire the full range of ammunition types, and the ordnance certainly fires HE and tactical nuclear projectiles to a range of about 30000 m (32,810 yards). It is probable that an accompanying vehicle carries most of the gun crew and the bulk of the ready-use ammunition supply.

Specification (provisional)
SO-203
Type: self-propelled 203-mm (8-in) gun with a crew of unrevealed size
Weight: 40 tonnes (39.37 tons)
Armament: one 203-mm (8-in) gun

with an unrevealed number of ready-use rounds on a powered mounting capable of elevation between an unrevealed depression and +60° and an unrevealed traverse angle
Performance: not revealed
Dimensions: length overall 12.8 m (42 ft); width 3.5 m (11 ft 6 in)

Powerplant: one 336-kW (450-hp) diesel engine of unrevealed type
User: Former USSR

210
GIAT GCT

FRANCE

Designed to replace the obsolete Creusot Loire Mk F3 equipment based on the chassis of the AMX-13 light tank, the **Grande Cadence de Tir** entered service in 1978 with Saudi Arabia, and then with France under the designation **155 AU F1**. This self-propelled 155-mm (6.1-in) gun is based on the well-proved chassis of the AMX-30 MBT for commonality of maintenance and performance with France's armoured formations, and locates the main armament in a substantial electro-hydraulically powered turret on the vehicle's centre of gravity.

The main armament has a 40-calibre barrel, and the installation of an automatic loading system and a 42-round supply of ready-use ammunition

provides a high rate of fire: a burst capability of six rounds in 45 seconds, compared with three rounds per minute with manual loading. The ammunition supply can be replenished with the GCT still in action, and the ordnance fires the standard varieties of ammunition (HE, submunition dispenser, smoke and illuminating) to a range of about 21200 m (23,185 yards), and two types of rocket-assisted HE to over 30000 m (32,810 yards).

Specification
GIAT GCT
Type: four-crew self-propelled 155-mm (6.1-in) gun
Weight: 42 tonnes (41.34 tons)
Armament: one 155-mm (6.1-in) gun

with 42 ready-use rounds on a mounting capable of elevation between −4° and +66° in a powered turret with 360° traverse, and one 12.7- or 7.62-mm (0.5- or 0.3-in) AA machine-gun with 800 or 2,050 rounds
Performance: maximum speed

60 km/h (37 mph); range 450 km (280 miles)
Dimensions: length overall 10.25 m (33 ft 7.5 in); width 3.15 m (10 ft 4 in)
Powerplant: one 537-kW (720-hp) Renault HS 110 multi-fuel engine
Users: France, Iraq and Saudi Arabia

Combat Comparison

Both the Abbott and the M110 saw service with the Royal Artillery, although the British Army has now replaced both with its home-grown gun, the excellent 155-mm AS90 featuring advanced fire control.

The Russians also use SP guns to support their armoured formations, but they often use them rather differently. Many of their guns will follow closely behind the lead tanks in an attack and will fire directly at any target that appears. This gives a very quick response, but low-velocity artillery guns are not very accurate over open sights and their armour is barely bulletproof. Perhaps if you have enough SPs you don't mind some of them being destroyed!

The future for SP artillery looks very interesting. The British Army now has its own 155-mm, called AS90. This is proving to be an excellent design, having all the requirements of range, accuracy and explosive force to command the modern battlefield.

However, the most important new piece of SP artillery to come on the scene is the MLRS (Multi-Launch Rocket System). This is made by the Americans, but is used by many other nations, such as Britain and Germany. It is a fantastic system that fires rockets loaded with submunitions such as bomblets and mines. Its destructive ability is staggering, and it will make massed tank attacks very unhealthy indeed for the attackers! It does have a distinctive 'signature' on firing – clouds of smoke and a trail from the rocket – but being an SP it can 'shoot and scoot' and will be long gone by the time an enemy fires back.

Self-propelled artillery has come a long way from pom-poms on lorries. Soon the huge, modern guns will all have onboard computers so that they can be spread out but can answer a call for fire within seconds, with devastating effect.

"Shells save lives," as the gunners say – SP artillery gets the shells there faster.

Above: The M109A1 is seen with its own resupply vehicle, the M992 Field Artillery Ammunition Support Vehicle. It carries an additional 93 rounds and 99 charges. Inset: The reality of life inside the gun. The M109 is relatively spacious in its turret. Smaller guns, such as the Abbot or the Soviet 2S1, are not so comfortable.

211

UNITED KINGDOM

FV 433 105-mm Abbot

At the end of the last war the standard British SP gun was the Sexton with its 25-pdr weapon. Various prototypes of other guns were trialled with either the 25-pdr gun or a 140-mm gun, but were dropped following the adoption by NATO of the 105-mm and 155-mm as standard calibres. In order to meet this, development work was done by Vickers and ROF, Nottingham, to produce a new weapon. The result was the FV 433 Abbot.

The Abbot is an adaptation of the FV 432 APC, and as such is of all-steel construction. The driver is seated on the left with the engine compartment to his right. The commander and gunner sit on the right of the turret, and the loader on the left. The commander is provided with a 7.62-mm anti-aircraft machine-gun. Abbot is fitted with an NBC filtration system and IR headlights, but the latter are no longer used.

The gun fires to a maximum range of 17 km and fires HE, smoke and illuminating rounds. 105-mm is an inadequate calibre for modern artillery; the shell is too small and does little damage. It is also fired by too weak a charge, resulting in too small a range. To this end, the British Army began replacement of Abbott with the AS90, a far superior weapon, in the 1990s.

Specification
FV 433 105-mm Abbot
Type: four-crew self-propelled 105-mm (4.13 in) gun
Weight: 16.5 tonnes (16.3 tons)
Armament: one 105-mm (4.13-in) Royal Ordnance gun with 40 rounds on a mounting capable of elevation

212

USA

Pacific Car and Foundry M110

Based on the same chassis as the M107 175-mm (6.89-in) gun, the **M110** self-propelled 203-mm (8-in) howitzer entered service in 1963 as replacement for the elderly M43 and M55 equipments. The type was designed for the delivery of long-range fire, and was therefore produced with a completely unprotected mounting on the rear of the vehicle, and a total ready-use ammunition supply of only two rounds. Most of the 13-man crew is carried in the accompanying M548 vehicle, which also accommodates a much larger supply of ammunition.

The 26.536-tonne (26.12-ton) M110 was fitted with a short 25-calibre barrel, sufficient to fire the HE projectile to 16825 m (18,400 yards); other projectiles were submunition dispenser, chemical and 10-kiloton nuclear. In 1977 an improved model was introduced as the **M110A1** with a 37-calibre barrel for greater projectile range. In this variant the greater weight of the new barrel is partially offset by reduced fuel capacity

but, even so, range and speed are reduced. The definitive version introduced in 1978 is the **M110A2** with a double-baffle muzzle brake and able to fire the HE projectile to 21305 m (23,300 yards) as well as the rocket-assisted HE and improved nuclear projectiles to 29100 m (31,825 yards).

Specification
Pacific Car and Foundry M110A2
Type: five-crew self-propelled 203-mm (8-in) howitzer
Weight: 28.35 tonnes (27.9 tons)
Armament: one 203-mm (8-in) howitzer with two ready-use rounds on a powered mounting capable of elevation between −2° and +65° and 60° traverse (30° left and right)
Performance: maximum speed 55 km/h (34 mph); range 523 km (325 miles)
Dimensions: length overall 10.731 m (35 ft 2.5 in); width 3.149 m (10 ft 4 in)
Powerplant: one 302-kW (405-hp) Detroit Diesel 8V-71T diesel engine
Users: Belgium, Greece, Iran, Israel, Italy, Japan, Jordan, Netherlands, Pakistan, Saudi Arabia, South Korea, Spain, Taiwan, Turkey, UK, USA and West Germany

from –5 degrees to +70 degrees; one L4A1 7.62-mm (0.30-in) commander's anti-aircraft gun
Performance: maximum road speed 47.5 km/h (30 mph); range 390 km (240 miles)
Dimensions: length 5.84 m (19 ft 2 in); width 2.64 m (8 ft 8 in); height 2.49 m (8 ft 2 in)
Powerplant: one 179 kW (240 bhp) Rolls-Royce 6-cylinder diesel
Users: UK and India

One immediate advantage of the Abbot is that it has a turret, and so provides the crew with small-arms and splinter protection as well as NBC and weatherproofing. The only weatherproofing on the M110 is a good jacket.

The 105-mm gun is verging on the useless. It has a maximum range of 17 km and the 105-mm shell weighs only 16 kg. The calibre has long been declared obsolete by NATO and has been replaced by 155-mm as standard.

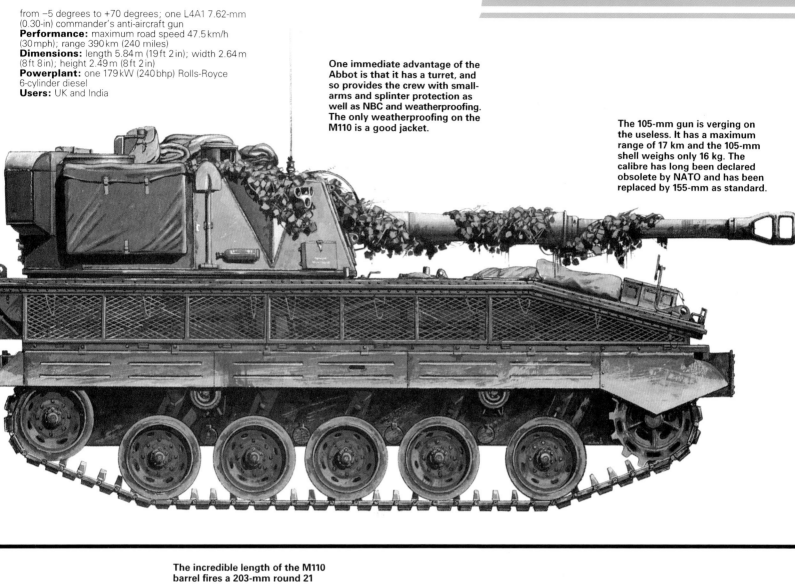

The incredible length of the M110 barrel fires a 203-mm round 21 km. It is nuclear-capable, unlike the Abbot. In terms of weight, a single HE round from the M110 is worth six from the Abbot. It is easy to see why the Abbot was withdrawn.

The lack of any creature comforts for the crew is far from popular with those who operate it. Local modifications have been seen of US Army M110s with tarpaulin covers over the back. They do nothing in the way of real protection, but probably make life a little more bearable.

HAVE GUN... WILL TRAVEL

A history of self-propelled artillery

The flexibility afforded by self-moving artillery was quickly realised, even during World War I. However, it was during the 1939-45 conflict that it came into its own.

The bombardment that preceded the dawn attack on the Somme was, they say, audible in London. In the static quagmire that the Great War soon became, the need for highly mobile artillery was minimal. The emphasis was on larger and yet larger guns that could wreak more and yet more destruction with each shell. The massive earth fortifications that scarred the plains of Northern Europe were disrupted only by guns like the German 'Big Bertha', the incredible 420-mm (16.54-in) howitzer that could fire an 810-kg shell over 9000 metres.

It was the next great conflict that really saw the consolidation of the self-propelled artillery concept.

The fast, hard-hitting and highly mobile tactics of Blitzkrieg in World War II called for equally fast-moving fire support. The Germans, although initially slow to fully implement this, were by the height of their campaign producing many outstanding weapons.

While the Western forces were toying with cardboard tanks and tankettes in the first part of the 20th century, Soviet planners, quick to realise the advantages of mobile artillery, were assembling huge armoured formations. Once the Nazi invasion of Russia had begun, followed by the uprooting of Soviet industry to the Urals, the luxury of diversification had to be foregone, so by 1945 the self-propelled artillery vehicles the Red Army took into the very heart of Germany were little more than a modification of a previous vehicle already in mass production – a trend still in evidence today.

Ancient principles

The last 40 years have seen a growth in sophistication of equipment, but no fundamental change. As electronics technology progresses, artillery computers become both faster and more accurate. Industrial production techniques mean much closer tolerances in mountings and sights, all of which means extra accuracy. But the principles, tactics and, to a large degree, the very nature of the beast are no different from those seen rumbling throughout Europe during the war. The similarities are greater than the differences.

World War I weapons

World War I was fought in the early years of motor transport, often using massively heavy vehicles and over the most difficult terrain. While it was realised that the new engines could be of great use, supply was still slow. But once the potential of the tracked gun carriage had been appreciated it was not long before experiments were in hand to harness the huge increase in mobility it afforded.

Not really a self-propelled gun at all, but so important to the Imperial German plans of World War I that it had to be mentioned, is the 420-mm 'Big Bertha'. This awesome weapon could fire an 810-kg shell into the heart of the fortified Belgian fortresses. The shell would penetrate deep in the earth before exploding.

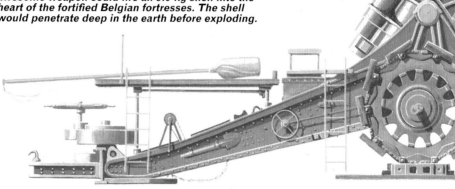

Above: The weapons that transformed the green, rolling countryside of northern France into this stinking quagmire may not have been self-propelled, but the effect was the same. The constant showers of high explosive destroyed everything they touched.

The biggest development has been in the rounds fired. Rocket-assisted projectiles greatly increase ranges. Terminal-guided munitions, such as Copperhead, turn artillery from an area weapon into a precision munition. Fuses now allow shells to explode in the air above troops, on impact with the ground, or can even delay detonation to explode in the ground, sending shock waves through the earth to disrupt trenches and destroy command centres.

Without a doubt, the biggest change that has occurred is the lethality the weapon affords. The shrinking in size of weapons of mass destruction (chemical, and especially nuclear, rounds) to artillery calibre has put in the barrels of the gunners a destructive ability vastly superior to anything in their history. Artillery has always been a battle-winning weapon, but today it can claim to be a weapon of Armageddon.

Above: An idea about to germinate. It does not need a massive leap of the imagination to go from dismantling your gun and transporting it on the Gun Carrying Tank to fixing one onto such a chassis permanently. The French made the leap by the end of the war; the British took rather longer.

Right: The railway gun was one method of overcoming the mechanical shortfalls of the time. The advantage of trains was that massive guns could be mounted on them and moved around with relative ease, within the obvious limitation of the railway tracks.

Range

Firing the conventional HE round, the M107, the M109A1 can deliver a 42.91-kg shell at a muzzle velocity of 683 metres per second out to a maximum range of 18.1 km. There is a rocket-assisted projectile (RAP) that has a maximum range of 24 km. Other rounds do not go as far. The HE(M731) round can deliver 36 anti-personnel mines out to 17.74 km, while the M485 illuminating round will only go as far as 13.6 km.

Air defence

For local air defence the M109A1 is fitted with the Browning .50 (12.7-mm) or, in some countries, the 7.62-mm (0.3 in) air defence gun. Its main function is air defence, although it serves as a local protection weapon as well.

Ammo

By virtue of its calibre, the M109A1 can fire most standard NATO ammunition. This includes conventional HE as well as proximity fused HE. In addition there are a number of mine, grenade or smoke-dispensing rounds as well as illumination rounds. Its most lethal ammo is either the M454 nuclear shell or the M121A1 VX or GB chemical round.

Spade

Every self-propelled gun needs an earth anchor and spade to assist in the absorbtion of the huge recoil. If the spade were not lowered the vehicle would rock violently when the round was fired. This would have two consequences. Firstly, it would cause inaccuracy because of the movement of the barrel. Secondly, it would slow the rate of fire since the crew would need to wait for the rocking to stop before firing again.

World War II weapons

Many of the early self-propelled platforms were simply conversions of existing tanks to mount artillery pieces. As the conflict progressed the sophistication of design improved, and more and more specialist vehicles were developed. Two distinct schools of thought emerged. One regarded self-propelled artillery as a simple add-on to existing artillery doctrines. The other regarded the mobile gun as a form of close-range direct-fire weapon to be used in close support. This gave birth to the so-called 'assault gun'.

This Heuschrecke carried a 105-mm field howitzer on a chassis based on existing tanks. The howitzer was removed from the chassis before firing.

Above: The Sturmgeschütz III was the classic assault gun of World War II, used to provide very close, direct fire support for the assaulting troops. In effect, it was a sort of halfway house between a tank and a true self-propelled gun. They would actually go in on the assault and be under the direct and immediate control of the commander. StuG III was armed with a 75-mm gun.

Above: Like many of the guns of the time, the Hümmel was based on a tank already in service, this time the PzKpfw III and IV. It had a 150-mm gun that fired a 43-kg round to a range of 13 km. It was a popular vehicle, being cheap and easy to manufacture. Its major drawback was a lack of ammo stowage – it carried only 18 rounds on board and so had to be constantly resupplied.

Right: The SdKfz 124 'Wespe' was a purpose-built carrier for a 105-mm howitzer, based on the chassis of the PzKpfw II light tank. It was first used during 1942 and had a crew of five.

Above: The US Army M7, known officially as the Carriage, Motor, 105-mm Howitzer, M7, was also used by the British, who nicknamed it the 'Priest' because of the 'pulpit' that housed the anti-aircraft gun. Amazingly enough, the M7 is still in service in some countries today!

Right: The Soviets were also proponents of the assault gun. The ISU-122 shown here had a powerful 122-mm gun that was fired at medium velocity to great effect. Although not numerically an important weapon, the ISU-122 pointed the way to future designs.

Fume extractor
The fume extractor reduces the amount of fumes and smoke that passes back into the turret once the round has fired. It works by drawing clean air into the barrel as the round passes. As the round continues up the barrel, more fresh air will be drawn in, thus clearing the barrel of smoke.

Muzzle brake
Right at the end of the muzzle is the muzzle brake. This reduces the recoil energy that the buffer absorbs. As the round fires, the barrel will be driven back into the turret. The muzzle brake directs some of the gases and blast backwards at the top of the muzzle, so counteracting the recoil force.

Crew
The M109A1 is manned by a crew of six – one commander, one driver and four crew. They will be responsible for loading the gun, setting the fuses, selecting the charges and laying and firing the gun. The driver is responsible for general maintenance of the vehicle, although the entire crew will assist in field repairs etc.

Turret
The turret is controlled by hydraulic pumps that enable the full traverse of 360 degrees. The gun mounting, the M127, can be elevated to a maximum elevation of 75 degrees and a mimimum depression of −3 degrees. On board stowage is for 34 rounds, although some countries (notably Israel) have increased this figure.

NBC protection
To enable the crew to operate in a nuclear, biological or chemical environment, there is a collective protection system. This works in two ways. Firstly, it has extremely fine filters that provide clean air for the crew to breathe. Secondly, it over-pressurises the hull, so forcing air out of any gaps and thus preventing the ingress of fumes, dust etc.

The US Army M109 has proved to be one of the most popular artillery pieces of modern times; over 4,000 of them have been sold for export in addition to the 3,500 in service with US forces. The original model, the M109, had a shorter barrel and consequently shorter range. This was replaced by the M185 barrel, which is 9 metres long. This added about 4 km to the maximum range. The resulting vehicle was designated M109A1. Since then, numerous modifications and updates have been introduced, the latest vehicle being designated M109A6.

The British Army operated 169 M109A1 and A2 models organised in Heavy Regiments. There were eight guns to a battery, and three batteries to a regiment. The regiments were divided amongst the various armoured brigades to provide some reinforcement of firepower. Some M109 regiments were grouped together to form general support regiments as part of the 1st Artillery Brigade.

M109A1 155-mm s

Ballistic sight
Every self-propelled gun has the ability to fire in the direct fire role. In theory this should never have to be used, but the M109 saw active service in just such a way in Vietnam. There they were firing a canister round – a shell filled with ball bearings that detonates just at the end of the barrel: a sort of 155-mm shotgun.

Hydraulic recoil buffer
The blast of the round going off creates a massive amount of energy that must be absorbed by the recoil mechanism. It is designed to absorb the energy quickly and under control by allowing the gun to run back in its mounting to its full recoil length and then return the gun to the firing position.

Engine
Under the armoured front decks lies the Detroit Motors turbocharged V-8 diesel. It is powerful enough to drive the 24-tonne vehicle at 56 km/h for a distance of 354 km. It can climb a gradient of 60 per cent and cope with a side slope of up to 40 per cent. The vehicle has a fuel capacity of 511 litres.

The British Bishop was an early attempt to produce a self-propelled gun by placing a 25-pdr on a Valentine tank chassis. The gun was mounted with fixed traverse and only limited elevation. It was not a great success.

Modern-day weapons

The great developments made by the end of World War II have been consolidated, but despite the increase in sophistication of vehicles, fire control computers etc, the self-propelled gun of today is fundamentally the beast of 50 years ago. The main difference has been in the growth of calibre. In the 1960s and 1970s, the 105mm was considered standard. This was then supplanted in the West by the 155mm. Soviet and Eastern forces had always gone for large-calibre guns. Today, the smallest self-propelled Russian weapon is 122mm; the largest, 203mm. It will be interesting to see whether any further expansion of calibre is possible or desirable in this format.

Right: These US Army nerve agent rounds are designed to be fired down 155-mm artillery guns, e.g. M109. The marking 'VX' indicates that the agent is thickened. The chemical has the consistency of Bostik glue and will give off a lethal odourless, colourless vapour for weeks.

Below: The French GIAT 155-mm GCT, a rather ungainly-looking weapon, is actually a well-designed gun. It has an autoloader that can fire at a rate of eight rounds a minute, firing a 43-kg shell out to a range of 21 km. GCT is in service with the French, Iraqi and Saudi Arabian armies.

Left: An M12 gun of World War II. Despite not being too sure what to do with them when they got them, the US Army found the M12 very useful. Its 155-mm gun was extremely effective and the vehicle was sufficiently powerful to keep up with fast-moving forces.

Left: The reduction in size of nuclear bombs from the massive 'Fat Man' dropped on Hiroshima down to a 155-mm shell has meant a vast increase in the destructive power of the gunners. This picture shows the obsolete US 280-mm gun, but the US Army has large stocks of the M454 nuclear round with W48 warhead, which can be fired by the M109 self-propelled howitzer as well as by the M198 towed artillery piece.

Right: For those who think that the gunners have an easy time, miles from the action, doing nothing, this is the reality of life on a gun. Firing at a rate of up to six rounds a minute means shifting 270 kg of shells. The explosion of the gun firing is ear-splitting, and the stink of the fumes pervades everything you eat and drink. And every time you move you have to dig trenches for the crew.

HAVE GUN ... WILL TRAVEL

Above: A little-known Soviet artillery piece was the 203-mm 2S7. It was introduced back in 1975 with scant photographic and textual information available. It was understood that it would be an army or front-level gun, and that a 2S7 regiment would be made up of three batteries of eight guns. The 203-mm projectile had a quoted range of 30km, but this was unconfirmed in the West.

Below: The adoption by the British Army of the Vickers AS 90 155-mm calibre gun as the replacement for the less than useful Abbot 105mm marked a dramatic increase in firepower. This is an excellent piece of equipment that, for once, has been thoroughly trialled and put through its paces before being adopted, rather than the other way round.

The Voice of the Guns:
Massacre at Suoi Da

In March 1967, Captain Randall T. Elliot was defending a fire support base near the Cambodian border. During Operation Junction City it was to become the scene of one of the most bitter battles of the whole Vietnam War.

"We were in very dangerous territory. We had mounted a major operation called 'Junction City'. We often saw the enemy, and our patrols heard them a lot at night: sure signs that a fight was brewing. They were watching and following us closely, waiting for a mistake. They must have wondered what we were doing there: it was so near the Ho Chi Minh Trail that they could bring major units against us rapidly."

There were no fixed front lines in Vietnam. Operations were carried out over wide areas of country, and were supported from centrally located artillery positions. Usually mounted on hills for maximum visibility and range, the fire support bases were protected by hastily erected defences and fortifications, and in high-risk zones would have an infantry force attached for further protection. Such fixed positions were like magnets to the North Vietnamese Army or to Main Force Viet Cong, and became the scenes of some of the fiercest battles of the whole war.

"We were assigned to a fire base with the usual artillery battalion. Another infantry battalion arrived, and the engineers constructed berms – walls – around the camp. They limited Charlie's ability to see us, and made his direct fire weapons less useful. But Jesus, it was scary out there. Even though we had pretty awesome firepower, and there were three battalions of us, we knew that they had at least three regiments in the region, outnumbering us three to one.

"At about 9.30 in the morning

Left: Crews of guns like this 175-mm M107 rarely saw their targets, but at Suoi Da they were firing over open sights at Viet Cong charging straight at them.

Right: A typical firebase held both towed and self-propelled artillery, with a perimeter track around which the protecting infantry might drive their APCs.

of 21 March 1967, we came under the most intense mortar and rocket bombardment I had ever experienced. Five, six, seven hundred heavy explosions rocked through our positions. That kind of firepower meant that something big was coming, and that we were the target. Then it went quiet. We knew that Charlie was coming.

Forming up

"You could see for almost a mile. There was virtually no underbrush in this area, but there were a few tall trees. We rushed to the berm. It was already hot, and the air was filled with dust and the smell of cordite. In the distance we saw them forming up, row after row, just like the British at the battle of New Orleans in the war of 1812. I called to my men to fix bayonets. A cheer went up from over 2,000 American throats. After long weeks, if not months, of hunting an elusive enemy, he was at last before us. And our guns."

Artillery has always been the 'King of Battle', accounting for more enemy casualties than any other arm, and the US Army deployed no fewer than 65 artillery battalions to Vietnam. The weapons they used ranged from lightweight 105-mm towed pieces through 155-mm towed and self-propelled howitzers to long-range 175-mm self-propelled guns and massive 203-mm self-propelled howitzers. An artillery battalion manning a fire support base had three batteries of between four and six guns or

Above: An infantry patrol sets out into the jungle during Operation Junction City. This was the largest operation up to that time, involving 22 American and four ARVN battalions in Tay Ninh Province.

Left: A self-propelled gun fires an H&I (Harassment and Interdiction) mission during Operation Junction City. This involved random fire into enemy-held areas, forcing the enemy to remain alert through the night and depriving him of sleep if nothing else.

Right: A shotgunner returns fire after an ambush. Patrols during Operation Junction City reported that the enemy was present in large numbers, and eager to fight.

howitzers apiece. Most of the time, they would be providing indirect support to infantry units operating many miles away, but occasionally the battle would come much closer to home.

Enemy approach

"We wanted to go over that berm and fight these people. Cooler heads prevailed, and we waited. Some of my men checked their ammunition and weapons, while others stared at the incredible sight unfolding before us. The first assault elements were only a football field's length away, running towards us. Some were firing and bullets began to hit around us. Sappers came into view, carrying charges to breach our berm. The fire base commander gave the order to open fire, and we were shooting before he'd finished the command.

"Our artillery pieces had been moved up to the berm, and were

firing beehive rounds as fast as they could. These were anti-personnel shells containing hundreds of small steel darts, and they ripped through the enemy lines, stitching enemy soldiers to the trees and nailing their rifles to their chests. Machine-guns and rifles firing fully automatic added to the cacophony as the infantry got in on the act.

"The North Vietnamese were coming from two sides. As they got closer, our men began to fall, and their mortar rounds kept coming into our position. A bullet ripped through my jacket, just under my left arm. In spite of the carnage wreaked by the guns, Charlie was coming through.

Air support

"They committed their reserves, and we felt the pressure. They were working their way around the firebase, and we were taking casualties. We were not yet desperate, but the going was getting harder. We called in air support, and after several strikes we sensed a change. They were about to quit. We couldn't hold back any longer, and with one loud scream and totally out of control, we went over the top. I couldn't remember if my pistol was loaded, so I picked up one of theirs.

"We attacked the few remaining units, acting as a rearguard. Gradually the smoke cleared and the firing stopped. We sent out a flying squad and they got the mortar crews and captured the mortars. The communists had been stupid to withdraw and leave their guns unprotected.

"The fight had lasted about 35 minutes. I began to count the

Above: Fire support bases were vitally important to American operations in Vietnam and, as such, were also prime Viet Cong and NVA targets. A single communist shell landing in an unprotected ammunition dump would devastate the entire base, so all ammunition was well dug in. Subsidiary dumps were also set up, to allow the firebase to keep on functioning if the main dump was hit.

Left: Mortars seemed out of place among the big guns of a fire support base. Nevertheless they were always present, as a vital part of the base defences. Mortars were good for dealing with enemy snipers, mortar positions and rocket launchers.

Right: An M110 eight-inch howitzer emits an ear-shattering burst on firing, its massive recoil digging the rear stabilising dozer blade deep into the fertile Vietnamese soil. The M110 was the largest mobile artillery piece used by the Americans in Vietnam, throwing a 203-mm shell weighing 90 kg over 16.8 km. Seventeen heavy artillery battalions were partially equipped with the M110, five in conjunction with 155-mm towed howitzers and 12 with the very long-range 175-mm self-propelled gun.

EYE WITNESS

"They came in waves. Maybe 200 at a time, across hundreds of yards of open space, right at the guns. All our guys had to do was load canister rounds, point the guns horizontally, and keep firing as fast as they could. Our fire cut through them like a reaper through corn. But as soon as the Reds were cut down, they would fall back and patch up their wounded before launching a fresh assault. A lot of the bodies we found after the battle had fresh bandages, so I guess they were coming for their second or third time. I tell you, the colonel said it was the hardest fighting he'd seen, and that included Korea and World War II."

Captain G. Shoemaker, Company Commander, 3rd Battalion, 4th Infantry Division

Above: The helicopter was vital to the success of fire support bases in Vietnam. It was used to ferry ammunition, personnel and supplies into remote hilltop locations far from any other means of communication.

Right: Fire support bases could be set up very quickly. Once set up they would be on call day and night, ready to support ground operations for miles around.

bodies I could see. There were over 200 dead, including 25 Americans. The final count was probably a hell of a lot higher. I looked at them. Small, young and dead. They never had a chance. They could not fight on our terms. But still they had come, wave after wave, into the teeth of direct artillery fire.

"A few hours later, a small bulldozer was flown in to dig a pit for the dead. I looked into the pit at the bodies, wondering at the dedication of these people. They would fight on empty stomachs

and with a handful of bullets against our massive firepower. I knew then that we could not win this war."

The battle for the firebase at Suoi Da was the worst communist defeat of the war to that time. American losses were 30 killed and 109 wounded. Four hundred and twenty-three communist bodies were recovered from the immediate area after the battle, but the total Viet Cong and North Vietnamese death toll must have been much higher. Infantry weapons and air attacks certainly

took a toll of the attackers, but it was the direct artillery fire that was credited with saving the firebase.

Prime targets

Firebases were to remain vital parts of the American military machine in Vietnam. As a result, they became prime Viet Cong

targets. Some were indeed to be overrun during the course of the war, but only by stealth and infiltration. Never again would the communists make the mistake of making human-wave attacks into the very muzzles of the guns. The battle at Suoi Da proved, once and for all, that using riflemen against artillery just does not work.

FIRE MISSION

Artillery is a battle-winning weapon. It is used to soften up a position prior to attack, screen off a position or even, with today's highly sophisticated control, destroy armoured vehicles.

The enemy lay in wait in their trenches, their observation posts alert. In the distance they heard the sound of armoured vehicles moving but, to their immediate front, there was nothing. What they didn't see was the artillery forward observation officer (FOO) crawling forward and watching their position. What they didn't hear was his radio message back to the fire direction centre (FDC) calling for the guns. They never even heard the M109s from 15 km away roar into action.

The first thing they were aware of was a regiment's worth of fire, all 24 barrels of high explosive, delay-fused detonation, 155-mm shells raining down on them for three minutes of fire. A total of nearly 10 tonnes of HE. Nothing survived.

1 The guns will be allocated in a number of ways. Usually the battle group that is seeing the most action will get the most assets. They may be allocated an entire regiment of 24 guns, at priority call, if need be. The usual allocation for the other battle groups would be more like one battery (eight guns).

2 Artillery fire means more than just lobbing a load of HE at anything that moves. It is a very precise, controlled operation. A typical concentration of fire for eight guns would be an area of only 150 metres square.

3 The quickest way of calling down fire is on pre-recorded targets. These are chosen by an FOO from a map and the co-ordinates are fed into the target computers. To call fire on one of these you simply tell the guns to fire on that target; all the ballistic data has been calculated in advance.

4 Even simple HE is not that simple. Electronic fuzes, known as controlled variable time (CVT) fuzes, detonate the shells in the air, showering lethal shrapnel splinters over a wide area. This is particularly effective against troops without overhead cover.

5 As well as HE, artillery can be used to give battlefield illumination. The 105-mm NATO illumination round provides light over a 350-metre radius for 30 seconds. The 155-mm round gives 75 seconds' worth over an area of two square km.

6 The other classic use of artillery is to provide a smokescreen. Three 105-mm Abbot guns could create a 600-metre screen for seven to eight minutes (weather permitting) with their onboard line ammunition holdings. With modern thermal imagers conventional smoke is not enough, and a new 'hot smoke' is being developed to defeat this method of observation.

7 One of the tactical problems with artillery is that once a gunline opens up it is very easy to locate. By using radar or sound-ranging it is easy to pin down the exact location, making it prone to counter-battery fire.

At the business end. The entire procedure of a fire mission is to get these guns firing where you want them to. Without the front end control, artillery fire is almost random and will do little more than make a lot of noise. Accurately controlled, it is deadly.

8 One of the most important tasks of artillery is to shell other gunlines. Known as counter-battery fire, a large number of batteries will be dedicated solely to this task in an attempt to try to limit the enemy superiority.

9 The Soviet Army had an unbelievably large number of artillery assets. With one barrel per 5 metres of frontage as a typical allocation for a breakthrough attack, they were the greats exponents of concentrated fire. They dedicated entire regiments solely to counter-battery fire.

2 Fire mission

Once the target is identified, the guns must be brought to bear on it. The information needed is fairly obvious. If you tell them nothing else, at least tell them the grid reference of the target. They need to know at what they are being asked to shoot. It makes a huge difference to the type of shells used if they have to take out a heavily dug-in enemy position, compared with a few soldiers advancing in the open. The guns also need to know how long to fire for. Don't forget, they could be up to 30 km away and have no idea what you are doing. For speed and accuracy all this information is put in a standard format: Where it is, what it is, what you want doing to it, and for how long.

Below: The job of bringing in a fire mission falls to the Forward Observation Officer. Germany has adapted the M113 to a specialist FOO vehicle with sophisticated sights, positional determination equipment and advanced communications.

Left: To call the guns in, the observer must know the grid of the target, the bearing to it from his position, and what the target is. This will be enough to bring the guns to bear and tell them the type of rounds to use. The FOO will also need to say when he wants fire to begin and for how long he wants the guns. This is a fire mission.

The track junction has been listed as a target

Pre-registered targets are much faster to fire since all the information has been calculated in advance. It is possible to adjust fire from these

target: 'one BMP infantry debussing Grid 871433, 840 degrees magnetic'

magnetic bearing to the target

observer

magnetic compass

grid reference of the target

grid

map

Below: Once the FOO has established the information needed, he will then request the number of guns he wants: from one battery up to several regiments if he thinks the target warrants it.

1 Target identified

As the battle group advances, the artillery commander will send forward observation parties out with the lead tanks, infantry or, very likely, recce. Their task is to get themselves in a position to observe enemy contacts and so bring down controlled and accurate fire. They have advanced communications that allow them to talk directly to the FDC, making response times fast. However, there will never be enough gunner parties for all the demands put on them, so every infantry and tank soldier is trained to bring in artillery fire.

One of the many uses for helicopters is to mount what is known as an Airborne Observation Post. From there the FOO will have an excellent view of his target. The only problem is that the helicopter must be well out of the way when the guns open up to avoid being hit during the shelling.

3 Allocating the guns

The requests for artillery fire will always outnumber the guns available. In the Falklands and Middle East wars it was found to be much better to have a large number of guns for a short space of time rather than a few guns pounding away for ages. It is the job of the FDC to allocate guns to targets. The real damage is done in the first few seconds of a barrage: once the rounds start falling, the enemy will take cover and the damage tails off dramatically. If you hit them massively hard and fast they have no time to take cover. In fact, artillery planning is now so finely tuned that in the next conflict, when the guns open up, every gun available will fire at once.

4 At the gunline

The business end of the game is the gunline, which can be literally miles from the action. In fact, if they are not miles from the action there is something very wrong — guns are not supposed to be in the thick of it. The guns are arranged in gunlines by battery, each consisting of six to eight guns. Despite the number of times they have to move, they will always try to dig the guns in for extra protection. The line is commanded by a gun position officer. His job is to take the target information from the FDC and convert it into the elevation and traverse for each gun. The degree of accuracy required is so fine that he has to allow for such things as how much the Earth will rotate in the time the shell is in the air!

Above and right: The Fire Direction Centre (FDC) will be responsible for co-ordinating the requests put in and matching these to the availability of the guns. They will know the state of the guns and their locations at all times and so will be able to prioritise the tasks. They may have to call higher-formation FDCs for additional support if needed.

Below and inset: What a gun line looks like in reality. This was a battery of M109s on exercise in West Germany. The vehicles are well spaced to minimise the damage in case the gun line is subjected to counter-battery fire. The cam nets do not disguise the fact that they are vehicles, but they do go some way to disguise the fact that they are guns.

5 Adjustment of fire

From putting in a request to rounds hailing down on the enemy takes but a few minutes. To the FOO on the ground all he wants to hear over his radio is "Shot, over" or "Splash". This tells him that the rounds are on the way. It can take anything up to 30 seconds from firing to the rounds hitting the ground. But his work is by no means over. After a flight of 30 km it is highly unlikely that the rounds will land exactly where he wants them to. His job is to radio back adjustments to get it spot on. Once they are, he will control the intensity and duration of the onslaught. Only when he orders "End of mission" can the gunline close down and prepare to bug out, and the FOO move on to the next target.

Below: The guns will initially fire adjusting rounds. These are single rounds that allow the FOO to make adjustments to hit the target. When on target he will radio "Rounds on target, fire for effect."

Right: The FOO will adjust for any sideways error first so that the only adjustment to be made is for range. This is a harder task. If he cannot get the exact distance, he will fire one round short of the target. The guns will add 100 metres. If this goes over the target, by dropping 50 metres the next round will be on target. This technique is called 'bracketing'.

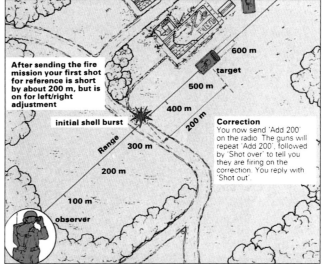

After sending the fire mission your first shot for reference is short by about 200 m, but is on for left/right adjustment

initial shell burst

Range

600 m

target

500 m

400 m

200 m

300 m

200 m

100 m

observer

Correction
You now send 'Add 200' on the radio. The guns will repeat 'Add 200', followed by 'Shot over' to tell you they are firing on the correction. You reply with 'Shot out'.

ROCKET ARTILLERY

Rocket artillery can deliver huge amount of firepower in a very short time. Thus it was very effective against the mass human waves used by the Iranians against Iraq in the war. The Iraqi BM-21 was used frequently, and to horrifying effect.

Able to fire massive quantities of high explosive over huge distances in seconds and then bug out before being hit by counter-battery fire, the multiple rocket-launcher is an incredibly powerful and destructive weapon.

It was a quiet, moonless night. The *kibbutzim* were asleep. It had been a hard day and the crop was still not all in. Twenty kilometres to the north and in another country, an ancient Soviet-made truck pulled to a stop. Its Palestinian occupants clambered out. Even in the darkness, it was obvious what it was. Twenty minutes after stopping, the quiet, clinging darkness was shattered. Forty high-explosive rockets smashed through the stillness and roared southwards. Seconds later they hit the *kibbutz*. The men, women and children asleep knew nothing of what happened. Over a quarter of a ton of high explosive obliterated the insubstantial buildings that had once been their homes. The lucky ones died.

The rocket has much to offer as a weapon. It is cheap and simple to build, yet when used *en masse* it can deliver huge amounts of explosives or other payloads extremely quickly. The BM-21 can fire its entire load of 40 rockets in just a few seconds. Perhaps its most frightening use is not as an area ex-

153

Above: Walid APCs of the Egyptian army parade through Cairo carrying 12-round rocket-launchers. These can fire a salvo of smoke rockets to create a smokescreen up to 1000 metres long, which can last for up to 15 minutes in favourable wind conditions.

Right: The Brazilians have a flourishing export arms market and have developed a wide array of rocket artillery systems. Shown here is the SS-30 rocket mounted on the Tectran 6x6 truck. The SS-30 has a range of 30 kilometres.

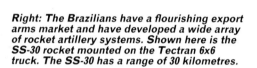

FLASHBACK

Rockets have been used for several hundred years. In World War I they were confined to signal flares, as employed here to warn of a gas attack. By World War II, they had become effective weapons.

More than fireworks

Rockets have been used in battle since the Chinese discovery of gunpowder about a thousand years ago. They were designed to affect enemy morale, creating bright flashes and unearthly noises.

Congreve introduced the first explosive rockets to the British Army during the Napoleonic wars. They were not very accurate, and at the battle of Leipzig they were as dangerous to their firers as to the French. More advanced weapons like the Hale spin-stabilised rocket were in use during the American Civil War, but accuracy was still poor.

This lack of accuracy meant that World War I rockets were used as signal devices. It was not until World War II, when more reliable propellants were developed and they were fired in barrages, that rockets finally became effective.

plosive weapon, but as a transport for chemical weapons.

Some types of chemical weapons, known as blood agents, require huge concentrations to be effective. With these agents being lighter than air, it is impossible to build up the concentration slowly using conventional artillery. The only way to deliver hydrogen cyanide is by mass volleys of rockets. This weapon can kill unprotected troops in less than a minute, yet disperses in under five minutes. Using rockets, it is possible to swamp an area, kill the defenders, and, in just a few minutes, drive through the area without the need of cumbersome chemical protective gear. Such is the power of the rocket.

Historically, however, the rocket has been

ROCKET ARTILLERY Reference File

255

FORMER USSR

BM-21 Grad

The standard multiple-launch rocket system of the USSR, Warsaw Pact countries and many of the USSR's other allies and clients, the **BM-21 Grad** is a truck-borne system that entered service in the early 1960s. The type has been built and adapted in other countries, while Russian developments include the **BM-21 M1975** and **BM-21 M1976,** which are respectively a 12-round version on a small 4×4 truck and a 36-round version on a ZIL-131 6×6 truck. The baseline type is usually allocated at the rate of one battalion of 18 launchers to each tank and motorised rifle division.

The BM-21 is based on the Ural-375D 6×6 truck, though the improved **RM-70** Czechoslovak version uses the

Tatra 813 8×8 armoured truck with a reload pack of 40 rockets. The launcher fires three types of rocket: a short rocket for a range of 11 km (6.84 miles), an augmented short rocket for a range of 17 km (10.56 miles) and a long rocket for a range of 20.38 km (12.66 miles). Warheads include 19.4-kg (42.8-lb) HE fragmentation, incendiary, smoke and chemical types, the last including a hydrogen cyanide variant. The BM-21 launcher generally travels with one reload pack, which can be loaded in 10 minutes.

Specification
BM-21 Grad
Type: seven-crew self-propelled 40-tube 122.4-mm (4.82-in) multiple-launch

rocket system
Dimensions: length overall 7.35 m (24 ft 1.4 in); width 2.69 m (8 ft 9.9 in); height 2.85m (9 ft 4.2 in)
Weight: 13300 kg (29,321 lb)
Traverse/elevation: 180° total (120° left and 60° right)/0° to +55°
Rate of fire: not revealed

Rocket range: 20.38 km (12.66 miles)
Users: Afghanistan, Algeria, Angola, Bulgaria, Chad, China, Cuba, East Germany, Egypt, Ethiopia, Hungary, India, Iran, Iraq, Israel, Libya, Morocco, Mozambique, Nicaragua, North Korea, Pakistan, Peru, Poland, Syria, Tanzania, USSR, Vietnam, Yemen and Zambia

Right: Overhead canopy rails are fitted to the South African Valkiri launcher as a camouflage to give the vehicle the appearance of an ordinary truck. The rocket's launch signature is minimal, which helps the system avoid counter-battery fire from the Soviet-era weapons it usually faces.

used as a conventional artillery piece. Technologically, the Germans were the most advanced of all World War II users, but they used rockets as an addition to conventional artillery, not as an alternative. It was the Soviets who were the first to realise their awesome effect and to deploy them as a weapon in their own right, and the Soviets are still the greatest proponents of rockets today.

The Germans deployed a large number of rockets of different size: from the small 15-cm Wurfgranate 41, the mainstay of the German arsenal, up to the 30-cm Wurfkörper 42. The larger calibres packed a bigger punch, but needed a larger motor to propel them. The trade-off between payload and propellant usually came down in favour of the smaller-calibre rocket, but vast numbers of these were required.

The Allies were not idle in rocket research, though. During the 1930s, the British had worked on a number of designs. Unlike the Germans, who used spin-stabilisation as a means of ensuring rocket accuracy, the British opted for fins. This was something of a poor choice. British rockets were wildly inaccurate. Slack mass-production with poor quality-control did nothing to help this. British rockets veered off their intended courses with alarming ease, often with dire consequences.

American attempts

The Americans, arriving rather late on the scene, were equally behind with their rockets. Their first attempts were based on British designs but they fared little better. Having captured a number of German launchers, they set about producing an accu-

rate spin-stabilised rocket, the M16. Unfortunately, no sooner had they perfected the rocket and launcher than the war ended, and the M16 was put on the shelf as yet another good idea, never really tried out.

As far as the West was concerned, the rocket rather fell from grace. It was used in the Korean War by the Americans, but not in great numbers, and was of indifferent consequence. In contrast, rockets were used widely by communist insurgents around the world. This has, until very recently, been the hallmark of the weapon.

The Soviets had in operation three rocket-launchers: the BM-21, BM-24 and BM-27. The BM-24 has been replaced, at least in the Russian Army, but nevertheless the system is still in use around the world, particularly in the Middle East. The Israelis captured a large

The Professional's View:

Multiple Launch Rocket System

"It is truly an incredible system. In terms of deliverable firepower, it is a massive increase in potential. If you add to that the fact that each launcher needs only three men to crew it, the destructive ability of a battery is such a huge step up, it's unbelievable. Of course, it has its problems. The weapon was designed for the Central Front war. If we take it that that war is not going to happen, then the system is not as flexible as conventional artillery. That being said, I'm sure we'll find a use for it. If you ask me which would I choose — MLRS or conventional tube artillery — I'd take MLRS any day."

Artillery Major, US Army, 1989

256

CHINA

Type 63 and Type 70 Launchers

The Chinese have indigenously designed and built two types of 19-tube 130-mm (5.12-in) calibre MRL systems: the **Type 63** mounted on the rear platform of the 2500-kg NJ-230 4×4 truck, in two variants; and the **Type 70** mounted on top of the Type YM531 tracked APC to replace elderly Soviet systems. The major difference between the two truck-mounted variants is that one has a covered crew cabin. All the launchers are used in batteries of six, the truck-mounted systems serving in MRL regiments and the APC types serving in the armoured divisions. The launch tubes are arranged in two rows, with a top row of 10 over a bottom row of nine. Both systems are in production and have

Type 70 MRLs serve with the armoured divisions of the People's Liberation Army. A similar system is mounted on trucks and used in the infantry divisions.

seen combat use with the Chinese and Vietnamese armies during their short border war in 1979. The North Korean army is also known to have the truck-mounted system in service, and may well be building it under licence in North Korean state arsenals as part of that country's arms-building programme.

Specification
Type 70
Type: six-crew 19-tube 130-mm (5.12-in) multiple-launch rocket system
Dimensions: length overall 5.48 m (17 ft 11 in); width 2.98 m (9 ft 9 in); height 2.63 m (8 ft 7 in)
Weight: 13.4 tonnes (13.18 tons)

Launcher traverse/elevation: 180° left and right/0° to +50°
Rate of fire: not known
Rocket range: 10.37 km (6.44 miles)
Users: People's Republic of China, North Korea and Vietnam

Weight of fire: rocket vs tube artillery

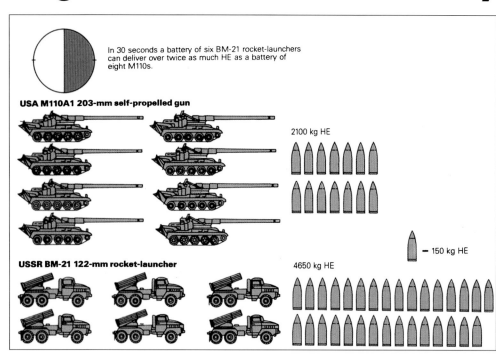

In 30 seconds a battery of six BM-21 rocket-launchers can deliver over twice as much HE as a battery of eight M110s.

USA M110A1 203-mm self-propelled gun

2100 kg HE

= 150 kg HE

USSR BM-21 122-mm rocket-launcher

4650 kg HE

The main tactical advantage of rocket artillery over conventional tube artillery is the sheer weight of fire that can be put down in such a short time. However, this is slightly misleading. Once BM-21s have fired, they must reload before they can fire their next salvo, whereas tube artillery like the M110 will just keep plugging away. The real point is that these two types of artillery complement each other; they are not interchangeable.

number of these during their numerous encounters with their neighbours and as a result have formed their own multiple rocket-launcher battalions, made exclusively from war booty. Their own 240-mm rocket is a marked improvement on the original Soviet one.

The BM-21, firing a 122-mm calibre rocket, remains the most numerous rocket-launcher in the world. It is a very simple system made up of a Ural-375D 6×6 truck with a 40-tube launcher mounted on the back. It is normally to be found in batteries of 18 in the divisional artillery group.

Latest Soviet equipment

The BM-27 is the most recent model. The rockets have been increased in calibre from 122 mm up to 220 mm, but the number on each launcher has been cut from 40 down to 16. As a combat weapon the BM27 had its baptism of fire when used against Israeli troops by Syrian forces in 1982, and by all accounts it is a powerful weapon. Its range is also greater than earlier Russian rocket launchers – some 40 km.

It is noticeable that if one looks through a list of producers of multiple-launch rocket

---

257

ISRAEL

Israel Military Industries MLRs

Israel is a major proponent of artillery used in combination with the multiple-launch rocket system. The former allows accurate engagement of point targets while the latter, cheap to produce and employ, can provide high volumes of fire for saturation tasks. The three Israeli-produced equipments are the **290-mm (11.42-in) Multiple Rocket System**, the **240-mm (9.45-in) Multiple Rocket System,** and the **160-mm (6.3-in) Light Artillery Rocket System.**

Entering service in the early 1970s, the 290-mm MRS is a massive weapon based on the chassis of the Centurion (initially Sherman) tank, and comprises a traversing and elevating launcher with four tubes (initially frames) for 600-kg

The 290-mm MRS is one of the largest systems in service.

(1,323-lb) rockets each carrying a 320-kg (706-lb) HE fragmentation or submunition warhead; four rockets can be power-reloaded by a single man in 10 minutes. The 240-mm MRS is really the Soviet BM-24: Israel has captured so many launcher vehicles that it now manufactures its own 110.5-kg (244-lb) rockets. The 160-mm LAR was designed to meet an Israeli requirement of 1974, but has been produced only for export. The rocket weighs 110 kg (242.5 lb) and carries a 50-kg (110-lb) HE fragmentation warhead over a range of 30 km (18.64 miles).

Specification
Israel Military Industries 290-mm Multiple Rocket System
Type: four-crew self-propelled four-tube 290-mm (11.42-in) multiple-launch rocket system
Dimensions: not revealed

Weight: 50,800 kg (111,995 lb)
Traverse/elevation: 360° total/0° to +60°
Rate of fire: not revealed
Range: 25 km (15.53 miles)
User: Israel

258

FORMER USSR

BM-27

Entering service in the mid-1970s to supplement the BM-21, the **BM-27** is a highly capable multiple-launch rocket system based on a ZIL-135 8×8 truck chassis. The type is allocated to Category 1 tank and motorised rifle divisions (one battalion of 18 launchers and 36 reload vehicles). There are also a single battalion in the artillery brigade of the combined arms army, a three-battalion regiment in the tank army, and an unknown number of battalions in the rocket-launcher brigade of each front's artillery division.

The launcher unit comprises one row of four tubes and two rows of six tubes. With the launcher traversed to a right angle with its vehicle, a full complement of rockets can be reloaded

The BM-27, known in Russian service as Uragzy, or Hurricane, has seen combat use in Afghanistan.

from the accompanying vehicle (based on the same chassis) in between 15 and 20 minutes: individual rockets are swung into position by a small crane and power-rammed into the relevant tube. The vehicle is stabilised in firing position by four hydraulically operated jacks (two at the rear and one on each side). The rocket weighs 360 kg (794 lb), has a minimum range of 5000 m (5,470 yards), and includes in its warhead options HE, chemical and submunition types.

Specification
BM-27
Type: six-crew self-propelled 16-tube 220-mm (8.66-in) multiple-launch rocket system
Dimensions: length overall 9.3 m (30 ft 6.1 in); width 2.8 m (9 ft 2.25 in); height 3.2 m (10 ft 6 in)

Weight: 22750 kg (50,154 lb)
Traverse/elevation: 240° total/0° to +55°
Rate of fire: not revealed
Range: 40 km (24.86 miles)
Users: Former USSR and Syria

Above: *Developed to provide heavy fire support to lightweight units, the 70-mm RADIRS is an adaptation of a standard American air-launched rocket pod. It can fire a rocket to over 15 kilometres and has a wide variety of warheads, including shaped-charge bomblets, smoke and illuminating rounds.*

Right: *The 70-mm SBAT-70 36-tube trailer-mounted rocket-launcher system is based on the standard Brazilian Avibras aircraft rocket, which has been adapted to a ground-based system.*

systems, a lot of Third World and former Warsaw Pact countries feature, yet few Western ones. One nation that has a thriving arms industry and is typical of multiple-launch rocket systems producers is Brazil. It manufactures a number of relatively cheap rocket-launchers, from the SBAT-70 70-mm trailer

---

259

WEST GERMANY

Wegmann LARS

The **Light Artillery Rocket System** was developed in the mid-1960s and taken into West German service during 1969. There is an eight-launcher battery (together with two Contraves Fieldguard fire control radars and a resupply vehicle with 144 rockets) to each division. The initial service standard was the LARS-1 on the Magirus-Deutz Jupiter truck, but during the mid-1980s this was upgraded to the considerably improved LARS-2 standard using a MAN 6×6 truck for better mobility, a new fire control system and more types of rocket warhead.

The launcher unit has powered traverse and elevation, and comprises a central pedestal flanked by two 18-tube banks of rockets. The 110-mm (4.33-in)

rocket weighs 35 kg (77 lb) and, in its original form, has minimum and maximum ranges of 6 and 14 km (3.73 and 8.7 miles); two later rocket types have improved propellants for maximum ranges of 19 and 25 km (11.81 and 15.53 miles) respectively. The rocket of the LARS-1 can carry HE fragmentation, submunition (including one with eight AT-1 anti-tank mines) and smoke warheads, while the LARS-2 type has improved versions of these warheads as well as adding radar-ranging and new submunition warheads, the latter with five AT-2 anti-tank mines. Reloading takes 15 minutes.

Specification
Wegmann LARS-2
Type: three-crew self-propelled 36-tube 110-mm (4.33-in) multiple-launch rocket system
Dimensions: length overall 8.28 m (27 ft 2 in); width 2.5 m (8 ft 2.4 in); height 2.99 m (9 ft 9.7 in)

Weight: 17480 kg (38,536 lb)
Traverse/elevation: 190° total (95° left and right)/0° to +55°
Rate of fire: 36 rockets in 17.5 seconds
Range: 25 km (15.53 miles)
User: Germany

260

BRAZIL

Avibras ASTROS II

The Brazilian company Avibras is one of the world's leading designers and manufacturers of multiple-launch rocket systems, mainly for the export market. The company's products range in rocket calibre from 70 mm (2.76 in) upwards, and the most versatile type is the **Artillery Saturation Rocket System II,** a modular system whose launcher, fire control and resupply units are all based on the TECTRAN 6×6 truck with four outrigger legs which are lowered for stability in operations. The fire control unit is based on the Contraves Fieldguard radar system, with each such unit controlling between four and eight launchers.

The fire unit can be fitted with any of three launchers: a 32-tube unit for the

127-mm (5-in) SS-30 rocket, a 16-tube unit for the 180-mm (7.09-in) SS-40 rocket, and a four-tube unit for the 300-mm (11.81-in) SS-60 rocket. The SS-30 weighs 68 kg (150 lb) and carries a 20-kg (44-lb) HE warhead over a range of between 9 and 30 km (5.59 and 18.64 miles), the SS-40 weighs 152 kg (335 lb) and carries a 54-kg (119-lb) HE or cluster warhead over a range of between 15 and 35 km (9.32 and 21.75 miles), and the SS-60 weighs 595 kg (1,312 lb) and carries a 160-kg (353-lb) HE or cluster warhead over a range of between 20 and 60 km (12.43 and 37.28 miles).

Specification
Avibras ASTROS II/SS-40
Type: three-crew self-propelled 16-tube 180-mm (7.09-in) multiple-launch rocket system
Dimensions: not revealed
Weight: not revealed

Ammunition: HE, submunitions
Traverse/elevation: not revealed
Rate of fire: not revealed
Range: 35 km (21.75 miles)
Users: Iraq, Libya and possibly others

A battery of Soviet-made BM-21s can deliver four and a half tons of chemical weapons in a single salvo

The internationally developed MLRS is currently in service in a large number of NATO countries. It is a truly fine artillery piece with much to offer, but at quite a cost.

up to the 300-mm Astros II, the latter being in service with the Iraqi army

The biggest Western development of the past decade has been the Vought MLRS (Multiple-Launch Rocket System). This 227-mm, 12-tube, high-mobility system is a fearful tool of broad-area destruction. The MLRS has been taken into service in many countries outside of the US, and during the Gulf War it inflicted appalling casualties upon the Iraqi forces exposed on the flat desert plains.

MLRS was designed for a rather different use than the Soviet weapons. Their rockets are exclusively area weapons, designed to swamp an area with explosives and simply obliterate

entire grid squares at a time. The MLRS is rather less crude. Its rockets are better made and much more accurate; its fire control system is a bit more than the 'light the blue touchpaper and retire to a safe distance' methods of the BM series.

MLRS delivers a variety of warheads, but they all operate on the submunitions principle. It is more effective to deliver thousands of little bomblets on a target than one almighty great bang. Bomblets will devastate soft-skin targets and have the additional advantage of denying an area, since unexploded ones form rather nasty anti-personnel mines. MLRS can also deliver anti-tank mines or anti-tank munitions.

It is rather difficult to predict the future for the rocket. Certainly the small and cheap Third World systems will continue to be developed and sold around the world; the Middle East seems a good place to set up shop. Prior to the Gulf War, the future of weapons such as the MLRS seemed questionable. Yet having seen the results it can impose on the battlefield, more and more nations are taking to the concept and placing orders. As a weapon of sheer attrition, the MLRS has few equals in land warfare.

In contrast to MLRS, the BM-21 is a cheap, mass-produced weapon. But what the weapon lacks in sophistication is more than made up for in numbers.

First to Last

The Soviet Katyusha was introduced in World War II. It was crude, yet cheap and simple to operate. It has proved to be an incredibly long-lived weapon, still in service today. In contrast, MLRS is the absolute state-of-the-art type. Fitted with electronic guidance and locating equipment, it is vastly more expensive and much more prone to breakdowns.

261
USSR

Katyusha

It is slightly misleading to give the name **Katyusha** to one particular weapon system. In fact the Soviets tried a variety of different rocket and truck combinations and called each of them Katyusha. The most common combination was the unlikely marriage of the Soviet M-13 132-mm rocket with a US-supplied Studebaker 6×6 truck.

The first battery went into operation on 7 July 1941 and was used against an advancing German troop concentration. The sudden deluge of rocket fire caused mass panic among the German troops who were unlucky enough to be the first recipients of the rain

262
USA

Vought Multiple Launch Rocket System

Though the US Army had not previously been a keen exponent of the multiple rocket-launcher concept (only two such systems have been fielded since World War II), in 1976 the Redstone Arsenal launched a feasibility study for a General Support Rocket System as a low-cost weapon offering a high rate of fire and capability against troops, light equipment, air defence systems and command posts. Five initial submissions were whittled down to two before trials started in 1979, and in 1980 the Vought **Multiple Launch Rocket**

Since MLRS vehicles are designed to operate as single units spread over a wide area, they are equipped not only with sophisticated VHF communications, but also with a secure digital link to the fire direction centre, which enables fire missions to be sent direct to the onboard computer.

Like Katyusha, MLRS has a similarly-tasked set of support vehicles, designed to keep the system in action. However, reloading of the vehicle is much simpler. MLRS rockets are delivered in their own containers, which are loaded directly into the vehicle via a built-in crane. Katyusha was reloaded entirely by hand.

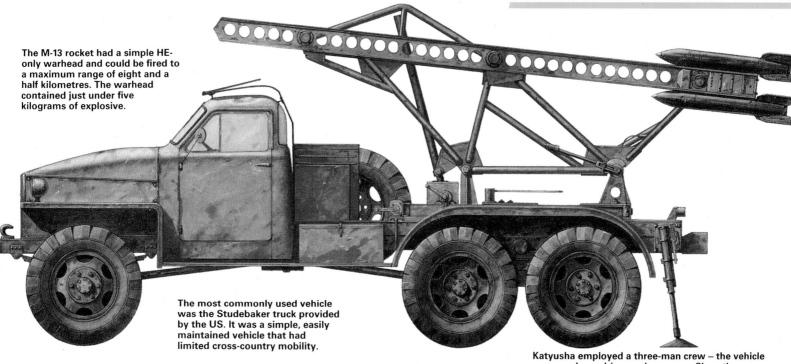

The M-13 rocket had a simple HE-only warhead and could be fired to a maximum range of eight and a half kilometres. The warhead contained just under five kilograms of explosive.

The most commonly used vehicle was the Studebaker truck provided by the US. It was a simple, easily maintained vehicle that had limited cross-country mobility.

Katyusha employed a three-man crew – the vehicle commander, a driver and a gunner. Since the vehicle was never intended to be operated on its own, but always as part of a battery, it had no communications or other equipment.

of high explosive and metal fragments. It was the dreadful scream of the incoming rockets that caused the Germans to nickname the Katyusha 'Stalin's Organ'.

When the Germans invaded the Soviet Union in 1941 there were only a few rockets and even fewer launchers. These first launchers were ZiS-6 6×6 trucks with rails for 16 rockets. So secret was the development of the weapon that only trusted members of the Communist Party were allowed as crew.

The M-13 rocket had a range of about 8000 metres and an HE fragmentation warhead. It was massively inaccurate, but since Katyushas were always used in mass numbers this mattered little.

Specification
Katyusha (M-13 on Studebaker US 6×6)
Type: three-crew 16-rail 132-mm (5.2-in) multiple-launch rocket system

Dimensions: length overall: 6.50 m (21 ft 4 in); width 2.24 m (7 ft 4 in); height 3.02 m (9 ft 11 in)
Weight: (vehicles only) 4558 kg (10,050 lbs)
Launcher traverse/elevation: not known
Rate of fire: all 16 rockets in under a minute
Rocket range: 8.50 km (5.28 miles)
Users: Soviet Union in World War II. Still in service in several Middle Eastern countries

System was declared winner for what had by then become the considerably more advanced NATO Multiple Launch Rocket System, which began to enter American service in 1982.

The firing unit is the Self-Propelled Launcher Loader, which is based on the chassis of the M2/M3 Bradley fighting vehicle series and carries on its rear a traversing and elevating launcher for 12 rockets in two six-round pods, which can be power-reloaded from a resupply vehicle in 10 minutes. The four rockets in service with modern forces are the Phase I US type

with 644 anti-tank and anti-personnel bomblets, the Phase II German type with 28 para-retarded anti-tank mines, the Phase III US type with six radar-guided anti-tank submunitions, and the Phase IV US type with a binary chemical warhead. Several other developments are in hand.

Specification
Vought MLRS
Type: three-crew self-propelled 12-tube 227-mm (8.94-in) multiple-launch rocket system
Dimensions: length overall 6.972 m (22 ft 10.5 in); width 2.972 m (9 ft 9 in); height 2.617 m (8 ft 7 in)
Weight: 25191 kg (55,535 lb)
Launcher traverse/elevation: 240°, though maximum elevation produces an overall height of 5.925 m (19 ft 5.25 in)
Rate of fire: not revealed
Rocket range: 30 km (18.64 miles) for Phase I rocket and 40 km (24.86 miles) for Phase II rocket
Users: France, Italy, Netherlands, Turkey, UK, USA and Germany

MLRS rockets are highly sophisticated multi-warheaded weapons dispensing a variety of submunitions, including anti-personnel and anti-tank mines. They have a range of over 30 kilometres.

Main picture and inset: The mighty 32-cm Wurfkörper had a 40-kg warhead. Its main problem was that it was too heavy and so had far too short a range. However, when sited to overcome this handicap, it was an awesome weapon.

THE SOUND AND THE FURY
Rocket Artillery in World War II

Reborn into the modern arsenal after nearly four decades of neglect, rocket artillery saw its apogee in World War II. It was used by all sides and proved itself numerous times.

The war rocket is an old weapon that was resurrected during World War II to supplement existing attack and defence systems (artillery and anti-aircraft weapons). The rocket has much to offer the weapon designer as it is a relatively cheap and simple device that can be mass-produced with comparative ease, and when used *en masse* is capable of fearful devastation. But in the term *en masse* lies the main failing of the war rocket: it has to be used in great numbers to

ensure that it will hit a precise target, for the rocket is inherently a projectile that will depart from a pre-selected trajectory with alarming ease and with little apparent reason. Set against this is the fact that it can carry a powerful payload for the costs involved, and so the arguments for and against such weapons continue.

The arguments for the rocket were in the ascendancy during World War II, nearly all the major protagonists making operational

use of them to some degree. Mainly it was to supplement existing weapons, but the Soviets discovered that the rocket could at times be regarded as a weapon in its own right. Technologically the Germans were the most advanced of all the World War II users, but they operated rockets as a supporting weapon to eke out artillery barrages, and only rarely attempted to deploy their rocket systems in the same offensive manner as the Red Army used its various Katyushas. The Katyushas were nearly always in the forefront of the offensive to oust the German invaders from the Soviet Union, and some of the Soviet rocket types used during those times are still in service all over the world. Indeed, the Red Army has maintained the multiple rocket-launcher as an important item in its military inventory, and has improved the performance of the relevant rockets to an extra-

ordinary degree. Only now are the Western nations beginning to re-learn the importance of this weapon as a counter to massed armoured and infantry attacks.

Rocket power

But during World War II the rocket made a considerable impact on many ground campaigns. The German *Nebelwerfer* units often tipped the balance in their application of heavy barrage fire during several battles, and, over the UK, British rocket batteries played their part in the defence of the nation. At lower tactical levels, the various American M8 weapons were often used to devastating effect in the reduction of strong-points and bunkers. Only the Japanese failed to use the rocket to its full effect. They did make some attempts in this direction, but for them the main problem was production, not tactics.

Left: The 15-cm Wurfgranate 41 was the mainstay of the German army Nebelwerfer units. It was designed as a smokescreen weapon, but was quickly re-assigned to its rocket artillery role. It had an unusual design in that the payload was behind the rocket motor. Thus when the rocket exploded, the additional shrapnel from the motor enhanced its lethality.

Below: The short-ranged but powerful 28-cm and 32-cm rockets were among the first to be fitted to vehicles, in this case the ubiquitous SdKfz 251. This conversion was known as the 'Foot Stuka' or 'Howling Cow'.

SdKfz 251 with 28-cm Wurfkörper

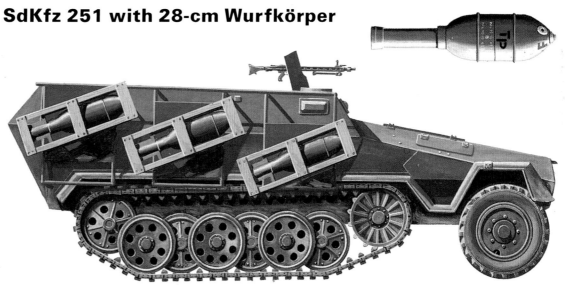

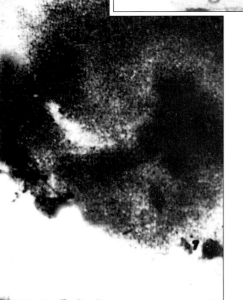

German rockets

The German rockets of World War II were organised into the *Nebelwerfer* brigades. The main sub-unit was the *Abteilung* (battalion). It had a headquarters plus two or three batteries. If more than one battery was equipped with 21-cm projectors it was known as a heavy battalion. The usual complement was two 15-cm batteries and one 21-cm battery. The Germans also deployed 28-, 30- and 32-cm projectors. These larger calibres suffered from shorter ranges, but if that could be overcome their destructive power was fearsome.

Left: Compared with the 28-cm and 32-cm rockets which preceded it, the 30-cm Wurfkörper 42 was a considerable improvement in design when it appeared on the artillery scene in 1942. Not only was it aerodynamically much smoother and cleaner, but it had a much higher propellant:payload ratio than any other German artillery rocket.

Right: First appearing in 1943 on the Eastern Front, the 21-cm Wurfgranate 42 was an immediate success. It was designed to replace the 15-cm Wurfgranate 41, and as such represented a marked improvement in lethality. The crew would load the five tubes and wire up the charges. They would then retire to a safe distance and fire the full unit electrically.

161

Vehicle engine

The carrier is powered by a 500-hp, turbocharged diesel with an automatic transmission. The powerpack assembly is located beneath and behind the carrier cab. It is connected through two drive shafts to a left and right final geared drive; this rotates a sprocket that moves the track. It gives the vehicle a range of 480 km and a maximum speed of 64 km/h.

Firing the rockets

Despite the awesome firepower delivered by the MLRS, the system is easy to maintain and operate. For a fire mission, MLRS needs only the target location and the latest meteorological data from external sources. The launcher's position is accurately plotted by the onboard position-determining system. The fire control computer does the rest.

Launching mechanism

On the back of the Armoured Vehicle Mounted Rocket Launcher is the launcher-loader module. This holds the rocket pods and allows them to be elevated and traversed. It consists of a turret assembly that rotates to 194 degrees left or right of the stowage position and can elevate to a maximum of 60 degrees. Fitted to this is the cage assembly, which accepts the rocket pods and links them to the fire control system.

Multiple Launch Ro

Fire control system
Once the rocket pod is fitted there is no requirement for the crew to leave the cab until they need to load. The fire control system monitors, co-ordinates and controls all assemblies during the launch and controls all elevation and traverse as well as the rate of fire and fusing.

Communications
Launcher communications equipment located in the cab includes standard VRC 353 VHF radio for primary crew communications. Mission commands are sent from the fire direction centre by digital radio and are electronically processed on board.

Carrier cab
The cab is designed to take the three-man crew and enables the rockets to be fired under armour. It is constructed of aluminimum armour plate for crew protection from small arms and splinters and has ballistic louvres that cover the windshield during launch, providing protection from rocket blast and heat.

Left: Mounted on the SdKfz 4/1 Maultier half-track is another example of the 15-cm Wurfgranate 41. The initial design had only a six-barrel mount, but after 1942 this was increased to 10 barrels for extra firepower. The vehicle could carry 10 rounds loaded, plus another 10 rounds.

Inset: The last development of the 30-cm Wurfkörper was to introduce a new propellant. One major problem with rocket artillery was its huge launch signature: once the rockets were fired, the crews could expect counter-bombardment to follow very quickly. Towards the end of 1942, a much cleaner propellant was used, producing less smoke and exhaust.

Above: By 1945, the Red Army was knocking on Hitler's door. For this task, the BM-13 Katyusha proved to be one of the most effective weapons around. It is no exaggeration to say that the BM-13 had established itself as the best rocket artillery piece in the world. It is still in service today.

Left: Shunted off the road and abandoned, this Katyusha has seen its day. Because of the distinctive moaning sound the missiles made in flight, the Germans dubbed the weapon 'Stalin's Organ'. It was probably the weapon the German forces feared most. A Katyusha barrage was devastating.

M-30 and M-31 300-mm Rockets

Entering service in 1942, the M-30 300-mm rocket carried almost six times as much explosive as the M-13 found on the Katyusha. However, its heavier payload reduced its range to under three kilometres. The first mobile launchers were introduced in 1944, but after 1945 both the M-30 and M-31 rockets were dropped because of their short ranges.

The Multiple Launch Rocket System (MLRS) is a highly mobile rocket-launching system, capable of delivering a high volume of indirect fire in a short time against critical, time-sensitive and high-value artillery targets. It is designed to supplement conventional artillery in its modern requirement of accurate placement of massive firepower over large impact areas at extended ranges. MLRS was introduced in the US Army in 1983 when a battery of nine launchers was delivered to the 1st Infantry Division. Since then, the US Army has constantly expanded its MLRS resources, and now the armies of European countries such as Great Britain, Italy, Germany and France include the MLRS within their artillery units.

Reloading
The advantage of using rocket pods is the speed and simplicity with which reloading takes place. The pods are delivered to the launcher or at a drop-off point. To replace the full load requires two pods and can be done by the crew in a few minutes.

Rockets
MLRS fires a basic 227-mm rocket to a maximum range of over 30 km. The rocket has a solid propulsion motor that allows four types of warhead: the M77 munition warhead for delivering submunitions such as mines etc; the practice warhead; the AT-2 mine warhead, a German development for delivering the AT-2 mine; and a terminal guidance warhead.

Crew
The vehicle is crewed by three men: driver, gunner and section-chief. This is a great improvement on conventional artillery, which requires considerably more men at the gun line.

Left: As the Soviet army entered Prague, they were cheered by the inhabitants. The city largely escaped damage since it was never bombed. However, the large amounts of rubble that can be seen are evidence of the artillery damage that was done. This was probably the last time the Soviets were cheered in Prague. Their later entry into the city was certainly less popular.

Below: A World War II-vintage Soviet rocket-launcher finds its way into Israeli hands via the PLO. The Israelis have captured so many of these weapons that they have their own batteries founded entirely on war booty.

Rockets of the Soviets

The Soviets, by the end of the war, were the world's greatest operators of rocket artillery. In fact their main weapon, the Katyusha, is still in service today. The development work was done in the late 1920s and was so secret that only trusted members of the Party were allowed to operate the weapon. During the course of the war, the calibre and destructive power of the rockets increased from the early 82-mm rocket to the final M-30 and M-31, each of which boasted a 300-mm warhead.

Rockets of the West

The earliest developments by the British in the rocket field were aimed at producing an anti-air rocket. However, the technology was soon switched to produce rocket artillery. A number of British designs were tried, but never in large numbers. The United States, in true American style, threw massive resources into the project and developed two basic systems: the ubiquitous 4.5-in M8 in a plethora of launchers, and the less common M16.

Main picture: The only Western weapon used in comparable numbers to the Katyusha was the American M8 4.5-in rocket-launcher, seen here in action in the Hurtgen forest, Germany, in November 1944. The gunners are reloading, surrounded by the smoke of the last volley. The M8 was an easily adaptable weapon and was mounted on a number of vehicles, including the M4 Sherman (inset).

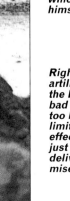

Left: The fire controller of the 1st Canadian Rocket Projector Unit looks towards German positions opposite 12 Corps, 15 January 1945. One of the problems of the early weapons was the huge backblast, which is why the gunner has dug himself in.

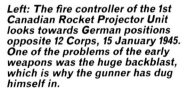

Right: The main British rocket artillery weapon was known as the Land Mattress. It was not a bad weapon, but was developed too late and deployed in too-limited numbers to have much effect. It had a maximum range of just over 7000 metres and could deliver a 7-lb payload with miserable accuracy.

KHARKOV:
prelude to Stalingrad

With a deafening roar, the rockets blast off towards the target, leaving behind thick clouds of smoke – a clear invitation to enemy artillerymen. Fired in such a massive salvo, the rockets will completely blanket the target position in the space of a few seconds.

Above: Row upon row of the huge 28/32-cm schweres Wurfgerät 41 rocket-launchers await firing. These were often placed in rows of four to achieve maximum saturation of the target area. The weapon was housed in a use-once-and-throw-away plywood frame that served as launcher as well as transport case.

Right: The 30-cm Wurfkörper 42 was introduced on the Eastern Front to beef up the existing artillery. It was technically a much better weapon, but for all the work done on it, its range was not much of an improvement on the weapons it replaced. It could fire its 44-kg payload over a maximum distance of 4500 metres.

In early 1942, the Germans were planning a campaign to defeat the Soviet Union before winter. The Soviets chose to attack at exactly the wrong time – as the Germans were massing for a major offensive.

The Germans had a field day. They managed to separate the Soviets from their artillery and so could fire without fear of the counter-bombardment they usually suffered.

The winter of 1941-2 shook the German army to the core. Not only was it the coldest winter for nearly 140 years, but the battles around the approaches to Moscow had led to defeat for the Germans and a shattering of the hopes and ambitions that had seemed so bright during the heady days of Operation Barbarossa in the early months after June 1941. As the winter days lengthened and the snows began to melt, the Germans found themselves deep within the Soviet Union and facing the prospect of a new summer campaign.

The Red Army was in little better shape during those early days of 1942. The Moscow battles had absorbed the last of their reserves from the Far East armies, and for once the Red Army leaders were forced to accept that they were running short of men. Materiel was also in short supply. The uprooted industries that had been installed east of the Urals were hardly functioning as yet and the Red Army had to face the prospect that none could be expected to produce weapons for months.

Even as the last battles in the snow were grinding to a halt, the staff planners on both sides were making their plans. Those of the Red Army were only too aware of the shortages ahead, and intended to make one last pre-emptive strike against the German army before the summer was far enough advanced for the Germans to make their expected moves. On the German side the high-level planning, conducted by Hitler himself, was less concerned with military gains than with economics. Hitler needed the oil

resources of the Caucasus region, for oil was fast becoming the weakest point of the German war economy, and without it the armed forces and industry would grind to a halt. The German army staffs were therefore given wide objectives to achieve, but for them the accent was on military success, and the oil question was hardly mentioned. Instead they planned a deep strike to the east and then north. The pivot of the northward strike, which was intended to reach beyond Moscow, was a city named Stalingrad, which was at that point only an interim objective. Following behind this massive move would be a deep strike to the Caucasus.

Pre-emptive strike

The *Nebelwerfer* units were to be very active in the months ahead, for even as the Germans organised their attacking forces the Soviets were pulling together what forces they could for a pre-emptive strike. Their objective was more modest than the German aims, for all they wanted to do was to recapture Kharkov, one of the major cities of the Soviet Union. In order to prepare the way and divert some of the German reserves, they started their campaign with a series of offensives in the approaches to the Crimea. These duly drew off the German reserves, but at a high price, for the Soviet moves were all stalled by an elastic defence. In the end the Red Army withdrew from the Crimea region altogether, leaving Sevastopol to a lengthy siege. The Soviets then turned their attentions to the area near Balakleya, where the fighting of January 1942 had left a salient in the German front.

Unfortunately for the Soviets, the German army also had its eyes on the region, which was to be the launching point for the Germans' eastward advance. Behind this area, therefore, the Germans were building up, under the aegis of Army Group 'South', two primary groupings to be unveiled in July. The main force was Army Group 'B' consisting of the 2nd, 4th Panzer and 6th Armies, the last commanded by General Paulus, a name to be heard much in the year to come. Army Group 'A' had only two major German armies, the 17th and 1st Panzer, the balance of its strength being made up of a collection of satellite forces such as the Hungarians, Romanians, Italians and others. Commanded by Generalfeldmarschall List, this army group was to move south

Below: One of the problems for the Germans became the speed of resupply. So confident were they of the Soviets' inability to reply that the batteries did not move, and ate up ammunition at an incredible rate. Their horse-drawn supplies were severely stretched.

into the Caucasus, with Baku as the objective for the spearhead formed by the 1st Panzer Army.

The unfortunate Red Army units were unwittingly moving right into the face of this massed array of force. Four Soviet formations (Group Bobrin and the 28th, 38th and 6th Armies of General Kostenko's South West Front) were ready to attack between Volchansk and Izyum about a week before the first German moves. Their idea was to sweep west, and then south and north to cut off the German forces around Kharkov. In typical Red Army fashion, their attack began on 12 May with a set-piece bombardment by artillery and massed Katyushas. The first attacks fell upon Romanian troops, who promptly broke and moved to the west as fast as they could. The Soviets poured after them in hordes, pursuing the hapless Romanians with horsed cavalry and T-34 tank support.

German trap

Unfortunately for them, the Soviets were advancing into a trap. As the Red Army tanks moved forward they were advancing into a gap between two of the major forces arranging for the German offensive: to their south was *Gruppe Kleist* (1st Panzer and 17th Armies) with all

its German divisions intact and ready to move, and to the north was the 6th Army, also ready to move. As the Soviet attack moved west, the German flanks held and the Red Army advance resembled a long thin pocket. Control of the Soviet forces was poor, for even as the threat to the westward advance was perceived the commanders were unable to do anything but continue the movement. Much of the Red Army's modern tank strength was committed to the attack, which simply had to work; it did not. On the morning of 17 May, the German forces on the southern side of the long pocket unleashed their countermove. All along the front, artillery and *Nebelwerfer* batteries heralded the advancing Panzers, which promptly cut across the neck of the Soviet advance and reached Izyum, the Red Army starting point. The bulk of the Soviet forces stretching to the west of Kharkov were trapped. Despite frantic efforts, the Soviets were unable to move in any direction and were contained in a number of isolated battles whose only perceptible feature was a large force attempting to move north towards Kharkov while the forces to its rear broke up into small isolated groups.

It was a field day for the

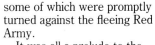

Above: A German rocket arches into the sky, kept roughly on course by its complex spin-stabilisation. Despite all the pressures of war, the Germans retained relatively sophisticated guidance systems, whereas the Soviets relied on simple fin-stabilisation – and big salvoes.

Below: The round of the 15-cm Wurfgranate 41 and its supply case are shown. This was the most numerous rocket of the war on the German side and was very effective. It was used in a high-explosive variety as well as in its original form as a smoke round.

some of which were promptly turned against the fleeing Red Army.

It was all a prelude to the coming summer campaign. In June the major German offensive duly opened and there was nothing to stop it. The losses of the Red Army in its ill-judged Kharkov offensive had sapped the last of its reserves and supplies of tanks. Beyond the Urals, the factories were working around the clock in atrocious conditions to produce more weapons to replace those losses, but trained men were in short supply. The way was thus open for the great move towards Stalingrad, the last of the Panzers' great sweeps, with the dreadful battles of the winter forgotten and replaced by a victorious rush across the steppes. To the south, the 1st Panzer Army swept into the Caucasus and much of the oil-producing region was taken.

But in less than a year it was all to change. Stalingrad, which in the winter of 1941 had been no more than a dot on the map, was to sap the last energies of the German forces in the east and the initiative was to pass to a newly enlivened Red Army.

Nebelwerfer batteries. With the Soviet units opposite them largely separated from their supporting artillery, the *Nebelwerfer* batteries were able to bombard their enemies with little risk of the usual counter-battery fire invited by the clouds of dust and dirt kicked up by their heavily-smoking rockets. Instead of the usual rush in and out of firing locations they could take their positions at an almost leisurely pace, set up their projectors, load them with their bulky rockets and take cover with less than the common haste. Once firing had started, accompanied by the usual cacophony of howling and roaring, the German crews could simply reload and prepare for the next massed salvo. To their rear, the half-tracks and trucks brought up more ammunition while their targets suffered and awaited the next onslaught.

End in sight

The end was not far off. On 19 May, Paulus nipped off the last major Red Army forces near Balakleya and the remains of the Soviet 6th, 57th and 9th Armies fled to the east, leaving all their heavy equipment behind in their rush to cross the Donets. They lost hundreds of thousands of men, all their precious tanks, and masses of artillery and Katyushas,

Smoke, heat and a huge dust cloud give away the launch of an MLRS salvo of 12 227-mm rockets. This is not a problem for MLRS, though. Having fired its salvo, it can be clear of the launch area in minutes - out of danger and ready for another fire mission.

The Multiple Launch Rocket System is a weapon of devastating firepower coupled with simplicity and speed. Never before have Western forces had such a versatile and useful weapon.

Rocket LAUNCH

Superiority in modern firepower requires accurate placement of massive firepower over large areas at extended ranges. An attacking force's most vulnerable area is immediately to the rear of its forward elements, where command and control, reinforcement, logistic and other supporting components are located.

At the same time, experience in the Falklands and other conflicts has shown that the maximum effect of any barrage occurs during its first few moments, before the target has time to take cover or evasive action.

With modern radars and sounding techniques it is possible to locate within a few minutes the exact location of a conventional artillery battery that has begun firing. Once

located, any battery is extremely vulnerable to counter-battery fire – it becomes an artillery target itself.

Looking at these three factors, one conclusion can be drawn about what you would want from an ideal artillery weapon – if such a thing were possible. It must be able to deliver massive amounts of HE very quickly, it must be able to hit targets behind the enemy's front line and it must be able to shoot-and-scoot, i.e. deliver its fireplan and then bug out. MLRS is just such a system.

MLRS has a maximum range of over 30 kilometres, and because of its sophisticated firing control system and advanced missile ballistics has an unprecedented accuracy at that range compared with other rocket artil-

lery systems. It can fire its full load of a dozen 227-mm missiles in under a minute. From arrival at its fire position, it can fire and be out of there again in less than five minutes.

However, MLRS is not quite the answer to every question. It is not appropriate for engaging single, isolated targets, its power being more suitable for dealing with larger areas. It is designed to complement, not replace, conventional tube artillery. Perhaps its biggest drawback is the reload time. Although for rocket artillery it is extremely quick, it still takes 10 minutes or so to reload. Conventional artillery can just keep banging away with its onboard ammo, while at the same time reloading. Operating together, as they are designed to, the two systems are unbeatable.

Above and left: Each launcher is designed to operate as an autonomous unit with its own pre-determined battle areas. It will wait in a hide for the 'go' order and then drive straight to its launch position, fire its rockets and bug out.

1 Deployment

Since each launcher is an autonomous unit, the idea of a gun line is unnecessary. The launchers will be split up into single units or perhaps pairs. They will deploy to battle hides such as woods, barns, farm buildings etc. From there, their pre-recce'd launch position will be only a few minutes away. Once the fire mission warning comes in, they will move out to the fire position and set up. When the order to launch comes through, the rockets will be fired and the vehicle will bug out to its next position. Compared with conventional artillery, this is much better use of terrain: with ordinary artillery, large areas of land have to be reserved for gun lines. But MLRS can share its hide with other vehicles, so making better use of the limited ground available in any war.

MLRS is also the basis for the US Army's new tactical missile system. This will use the same launch units to fire a new, much larger missile. With a range in excess of 100 kilometres, it is designed to take out enemy armoured formations, logistics bases and communications centres well back from the line of battle. It is designed to be fired using off-axis guidance techniques. Missiles are not launched in the direction of the target, but will change direction in flight. This is intended to prevent enemy radar systems from plotting the trajectory of the missile, and back-tracking to the launch position. This prevents the remote possibility of a quick-reacting enemy laying down counter-battery fire onto the firing unit as it leaves the launch position.

2 Resupply

Having fired its load of 12 missiles, MLRS must reload. This is a very simple operation. The launcher will drive to a reloading point where its next supply of rocket pods will have been pre-dropped. The rocket pod is fully integrated and requires nothing more in the way of charges etc to fire. The cage assembly, part of the launcher-loader module, is a welded aluminium structure. It aligns, holds and protects the rocket pods. But it also holds two boom-and-hoist structures that are used to pick up the new pods and feed them in.

Left: Reloading is a quick and simple operation. Each launcher has its own built-in crane. The rockets themselves come in pre-loaded containers or pods, six rockets to a pod. No extra equipment or manpower is required to fit the pods. They can be dropped in pre-determined positions so the launcher, having fired one mission, can go straight to a drop-off point, reload and drive to the next position ready for action.

3 Command and Control

With the launchers spread over such a wide area and operating as single vehicles, command and control is something of a problem. Although the vehicles are all connected by radio to the headquarters, this is of no use during periods of radio silence, i.e. when no transmissions are allowed. There is also the problem of crew fatigue. With only three men in a crew, operating 24-hour manning will be extremely tiring, resulting in a consequent decline in efficiency. This is a major problem which, along with control considerations, may result in a decision for the vehicles to operate in pairs.

Each vehicle contains its own onboard computer, which is linked by a secure digital radio link to the fire direction centre. It will receive the fire missions and calculate all the necessary ballistic data.

4 Fire missions

The order to fire and all the target data are sent to each launcher by a secure digital data link. This is a special coded message that cannot be intercepted by the enemy. Each launcher has its own onboard fire control computer and navigation system. Firing data is constantly updated, and the launcher is automatically laid and relaid onto target. The computer has an integral location system that constantly updates its own position so that the vehicle knows where it is at all times. To fire, all the MLRS needs to know is where the target is and some meterological data. By having its own onboard computer, MLRS can be programmed with fire plans well in advance, allowing an unprecedented amount of forward planning and so maximum use of the always-limited assets.

Above: Having arrived at their launch position, the crew do not need to leave the vehicle to launch the rockets; everything can be done from inside the cab.

Below: Since the MLRS established its credential in the Gulf War, investment has poured into th development of new warhead types. The large tactical missile warhead can been seen in the foreground here, alongside anti-materiel and anti-tank warheads.

5 Munition type

The MLRS has a range of warheads available, all with impressive area-denial effects. The main round is the DPICM rocket warhead (Dual Purpose Improved Conventional Munition) on the M77 rocket. A single rocket carries 644 anti-personnel, anti-materiel bomblets. The bomblet has a greater lethality than an equivalent weight in high explosive. A salvo of 12 rockets from a single launcher can place 7,728 bomblets on target in less than a minute, the equivalent of 22 M109s firing four rounds at the same time.

The Germans have developed the AT2 anti-tank mine for delivery by MLRS. One rocket contains 28 mines which are dispersed over an area of 100×100 metres.

The MLRS will also be used to launch the new US Army Tactical Missile System (ATACMS) which has a range of 100 kilometres. Two of these will be carried on each launcher.

In minutes, MLRS can lay an anti-tank and anti-personnel minefield over an area of several hundred metres at a distance of 30km. MLRS is a great asset of the modern arsenal.